Research Reports in Physics

Research Reports in Physics

Nuclear Structure of the Zirconium Region
Editors: J. Eberth, R. A. Meyer, and K. Sistemich

Ecodynamics Contributions to Theoretical Ecology
Editors: W. Wolff, C.-J. Soeder, and F. R. Drepper

Nonlinear Waves 1 Dynamics and Evolution
Editors: A. V. Gaponov-Grekhov, M. I. Rabinovich, and J. Engelbrecht

Nonlinear Waves 2 Dynamics and Evolution
Editors: A. V. Gaponov-Grekhov, M. I. Rabinovich, and J. Engelbrecht

Nonlinear Waves 3 Physics and Astrophysics
Editors: A. V. Gaponov-Grekhov, M. I. Rabinovich, and J. Engelbrecht

Nuclear Astrophysics   Editors: M. Lozano, M. I. Gallardo, and J. M. Arias

Optimized LCAO Method and the Electronic Structure of Extended Systems
By H. Eschrig

Nonlinear Waves in Active Media   Editor: J. Engelbrecht

Problems of Modern Quantum Field Theory
Editors: A. A. Belavin, A. U. Klimyk, and A. B. Zamolodchikov

Fluctuational Superconductivity of Magnetic Systems
By M. A. Savchenko and A. V. Stefanovich

Nonlinear Evolution Equations and Dynamical Systems
Editors: S. Carillo and O. Ragnisco

Nonlinear Physics   Editors: Gu Chaohao, Li Yishen, and Tu Guizhang

Nonlinear Waves in Waveguides with Stratification   By S. B. Leble

Quark-Gluon Plasma   Editors: B. Sinha, S. Pal, and S. Raha

Symmetries and Singularity Structures
Integrability and Chaos in Nonlinear Dynamical Systems
Editors: M. Lakshmanan and M. Daniel

Modeling Air-Lake Interaction Physical Background
Editor: S. S. Zilitinkevich

Nonlinear Evolution Equations and Dynamical Systems NEEDS '90
Editors: V. G. Makhankov and O. K. Pashaev

Solitons and Chaos   Editors: I. Antoniou and J. F. Lambert

Electron-Electron Correlation Effects in Low-Dimensional Conductors and Superconductors   Editors: A. A. Ovchinnikov and I. I. Ukrainskii

Signal Transduction in Photoreceptor Cells
Editors: P. A. Hargrave, K. P. Hofmann, and U. B. Kaupp

Nuclear Physics at the Borderlines
Editors: J. M. Arias, M. I. Gallardo, M. Lozano

Nonlinear Waves in Inhomogeneous and Hereditary Media
By A. A. Lokshin and E. A. Sagomonyan

A.A. Lokshin   E.A. Sagomonyan

Nonlinear Waves in Inhomogeneous and Hereditary Media

With 10 Figures

Springer-Verlag
Berlin Heidelberg New York London Paris
Tokyo Hong Kong Barcelona Budapest

Dr. Alexandr A. Lokshin
Research Institute "GEOINFORMSISTEM", Varshavskoye shosse 8,
Moscow, Russia

Dr. Elena A. Sagomonyan
Department of Mathematics and Mechanics, Moscow State University, Lenin Hills,
117192 Moscow, Russia

ISBN 3-540-54536-0 Springer-Verlag Berlin Heidelberg New York
ISBN 0-387-54536-0 Springer-Verlag New York Berlin Heidelberg

This work is subject to copyright. All rights are reserved, whether the whole or part of the material is concerned, specifically the rights of translation, reprinting, reuse of illustrations, recitation, broadcasting, reproduction on microfilm or in any other way, and storage in data banks. Duplication of this publication or parts thereof is permitted only under the provisions of the German Copyright Law of September 9, 1965, in its current version, and permission for use must always be obtained from Springer-Verlag. Violations are liable for prosecution under the German Copyright Law.

© Springer-Verlag Berlin Heidelberg 1992
Printed in Germany

The use of general descriptive names, registered names, trademarks, etc. in this publication does not imply, even in the absence of a specific statement, that such names are exempt from the relevant protective laws and regulations and therefore free for general use.

Typesetting: Camera-ready by authors

57/3140 - 5 4 3 2 1 0 - Printed on acid-free paper

Preface

This booklet presents a study of one-dimensional waves in solids which can be modelled by nonlinear wave equations of different types. The factorization method is the main tool in this analysis. It allows for an exact or at least asymptotic decomposition of the wave(s) under consideration in terms of first order multipliers.

Chapter 1 provides a general introduction. It presents some well-known results on characteristics, Riemann invariants, simple waves, etc. The main result of Chap. 1 is Theorem 1.3.2. (Sect. 1.3.2) which establishes the possibility of exact factorization of the nonlinear wave equation

$$\frac{\partial^2 a(\sigma)}{\partial t^2} - \frac{1}{\varrho}\frac{\partial^2 \sigma}{\partial x^2} = 0$$

with constant coefficients. This theorem permits one to construct further factorizations of more complicated wave equations which the reader will meet in the following chapters.

Chapter 2 is devoted to short wave processes in inhomogeneous media, the main result being the uniform asymptotic factorization of nonlinear wave equations with variable coefficients and the description of corresponding single-wave processes without the usual assumption of a small wave amplitude.

Chapter 3 deals with waves in nonlinear hereditary media. Nonlinear wave equations with memory that describe propagation of these waves can also be factorized (Sects. 3.2.1, 3.7.2, 3.7.3). One can see from the resulting single-wave equations that in the media under consideration two opposed factors come into play: nonlinear steepening of wave fronts and relaxation. As a result of this interaction, under certain conditions waves of stationary profile may arise, i.e., waves travelling in space without any change in shape. It is shown that such waves can be described by the nonzero solutions of homogeneous (Sect. 3.3) or self-coordinated (Sect. 3.4) nonlinear integral equations. From the purely mathematical point of view the existence of such waves solutions is not evident.

The results of Sect. 3.4 answer the question whether strong shock waves can propagate in nonlinear media with singular memory (i.e., a memory function which has an integrable singularity at a given time). This question is a natural consequence of the rigorous mathematical proof that strong shocks cannot propagate in *linear* media with singular memory. We use waves of stationary profile to demonstrate that for large enough wave amplitudes, the effect of nonlinear steepening turns out to be stronger than relaxation implying the possible existence of strong shocks even in the case of singular memory.

Another important result of Chap. 3 is the generalization of Landau's asymptotic approach to the investigation of elastic short waves of small amplitudes (Sect. 3.6).

The authors are grateful to N.V. Zvolinsky whose criticism helped us to improve these notes. The authors would also like to express their gratitude to V.M. Babich, V.L. Berdichevsky, M.A. Grinfeld, M.A. Itskovits, S.L. Lopatnikov, V.E. Rok, O.V. Semenenko, V.A. Shachnev and O.S. Vinogradova for useful discussions.

Moscow, 1988
Alexandr A. Lokshin
Elena A. Sagomonyan

Remark on the English edition

In the English edition of this book, Chapter 2 has been completely rewritten and now includes a discussion of nonlinearities of the general type. Results obtained since 1988, which illustrate the application of factorization theorems, are incorporated.

The authors would like to thank L.V. Demidova and O.I. Shuvaeva for the translation.

Moscow, October 1991
Alexandr A. Lokshin
Elena A. Sagomonyan

Contents

1. **Nonlinear Waves in Homogeneous Media** 1
   1.1 Preliminaries 1
       1.1.1 Equations of Motion of a Homogeneous Nonlinear Rod 1
       1.1.2 Riemann Invariants and Characteristics 2
       1.1.3 Simple Wave Equation 3
       1.1.4 Conditions on the Strong Shock 4
       1.1.5 Stability Condition for the Strong Shock 5
       1.1.6 Weak Shocks 7
   1.2 Nonlinear Hyperbolic Equations of the First Order 7
       1.2.1 Conditions on the Shock 7
       1.2.2 Constancy of the Integrals of Solutions 8
       1.2.3 Solution of the Boundary Value Problem Method of Characteristics 9
       1.2.4 Wave Breaking 11
       1.2.5 Principle of Equal Areas 12
       1.2.6 An Example 13
       1.2.7 Ordinary Differential Equation for a Shock Propagating into an Undisturbed Domain 14
   1.3 Exact Factorization of the Nonlinear Wave Equation with Constant Coefficients 15
       1.3.1 Introductory Observations 15
       1.3.2 Factorization Theorem for the Wave Equation for Stress 16
       1.3.3 Difference Between Linear and Nonlinear Factorization 17
       1.3.4 Factorization Theorem for the Deformation Wave Equation 18
       1.3.5 Earnshaw's Theorem 18
       1.3.6 Generalization of Earnshaw's Theorem 19
       1.3.7 A Boundary Value Problem Posed in Terms of Displacements 20
   1.4 Shock-Wave in a Simple System 22
       1.4.1 Formulation of the Problem 22
       1.4.2 Nonconformity of the Single-Wave Equation to the Shock Condition 23

		1.4.3	Transformation of the Single-Wave Equation. Integral Equation for $g(\sigma)$ Generating the Transformation	24
		1.4.4	Construction of the Function $g(\sigma)$	25
		1.4.5	Discussion of the Results	27
	1.5		The Shock-Wave in a Simple System (Continuation)	30
		1.5.1	Application of the Principle of Equal Areas	30
		1.5.2	Application of Euler's Method	31

2. Nonlinear Short Waves of Finite Amplitude in Inhomogeneous Media ... 33

   2.1 Asymptotic Factorization of the Nonlinear Wave Equation with a Variable Coefficient ... 33

      2.1.1 Representation of the Nonlinear Wave Equation with a Variable Coefficient ... 33

      2.1.2 Formulation of the Boundary Value Problem. Conditions of Asymptotic Factorization ... 37

      2.1.3 Single-Wave Solution of the Boundary Value Problem ... 38

   2.2 When is the Factorization Exact? ... 39

      2.2.1 Nonlinear Case ... 39

      2.2.2 Linear Case ... 40

   2.3 Asymptotic Factorization of the General Nonlinear Wave Equation with Variable Coefficients ... 42

      2.3.1 Preliminary Notes ... 42

      2.3.2 Notation ... 42

      2.3.3 Representation of the General Nonlinear Wave Equation with Variable Coefficients ... 43

      2.3.4 Formulation of the Boundary Value Problem Conditions of Asymptotic Factorization ... 46

      2.3.5 Linear Case ... 47

   2.4 Evolution of Maximal Amplitude of the Stress Wave ... 48

      2.4.1 Formulation of the Problem ... 48

      2.4.2 Equation for Maximal Amplitudes ... 48

      2.4.3 The Curve of Maximums as a Characteristic ... 49

   2.5 Propagation of a Stress Wave in a Homogeneous Nonlinear Elastic Rod Located in the Gravity Field ... 50

      2.5.1 Formulation of the Problem ... 50

      2.5.2 Uselessness of Exact Factorization ... 51

      2.5.3 Asymptotic Factorization ... 52

      2.5.4 Single-Wave Solution of the Problem ... 53

3. Nonlinear Waves in Media with Memory 55
3.1 Hereditary Elasticity 55
3.1.1 Linear Equations 55
3.1.2 Nonlinear Equations 56
3.2 Small Quadratic Nonlinearity 58
3.2.1 Asymptotic Factorization of the Nonlinear Wave Equation with Memory 58
3.2.2 Why Can't the Factorization be Exact? 60
3.2.3 Single-Wave Equation 60
3.2.4 Condition on the Shock for the Stress Wave 61
3.2.5 New Notation 62
3.3 Continuous Stationary Profile Waves and Nonzero Solutions of Homogeneous Integral Volterra Equations 63
3.3.1 Waves Propagating in an Undisturbed Medium 63
3.3.2 Integral Equation for the Wave of Stationary Profile 63
3.3.3 Estimate of the Solution of the Integral Equation .. 64
3.3.4 Existence of Stationary Profile Waves. Special Case 66
3.3.5 Existence of the Wave of Stationary Profile. General Case 69
3.3.6 The Exponential Kernel 71
3.3.7 The Simplest Oscillatory Kernel 72
3.3.8 A More Complicated Oscillatory Kernel 73
3.3.9 Waves Propagating in a Prestressed Medium 75
3.3.10 The Exponential Kernel 77
3.4 Stationary Profile Shock-Waves and Self-Coordinated Integral Volterra Equations 80
3.4.1 Waves Propagating in an Undisturbed Medium 80
3.4.2 Integral Equation for Stationary Profile Waves 81
3.4.3 Estimate of the Solution of the Integral Equation .. 82
3.4.4 Existence of Stationary Profile Shock-Waves 83
3.4.5 The Power Kernel 85
3.4.6 The Exponential Kernel 86
3.4.7 Waves Propagating in a Prestressed Medium 87
3.5 Waves Tending to a Stationary Profile 89
3.5.1 Intuitive Approach 89
3.5.2 Rok's Method 92
3.6 Nonstationary Waves Analog of the Landau-Whitham Formula 93
3.6.1 Formulation of the Problem 93
3.6.2 Linear Case 93
3.6.3 Case of Small Quadratic Nonlinearity 94
3.6.4 Estimate of Quality of the Approximate Solution .. 95
3.6.5 Single-Wave Equation for Deformation 96

	3.6.6	Single-Wave Equation for Displacement	97
	3.6.7	A Boundary Value Problem Posed in Terms of Displacement	98
3.7	General Nonlinearity. Further Factorization Theorems for Nonlinear Wave Equations with Memory		99
	3.7.1	Preliminary Notes	99
	3.7.2	The Exact Factorization Theorem	99
	3.7.3	The Asymptotic Factorization Theorem	101
	3.7.4	Waves in Rods in the Presence of External Friction	104
3.8	Nonstationary Waves for an Exponential Memory Function		105
	3.8.1	Formulation of the Problem	105
	3.8.2	Derivation of a Single-Wave Differential Equation	106
	3.8.3	The Analytic Solution in a Smoothness Domain	106
	3.8.4	Wave Breaking	107
	3.8.5	Case of Small Amplitudes. Asymptotic Analysis of the Shock-Wave	109
3.9	Reflection of a Wave from the Boundary Between Linear Elastic and Nonlinear Hereditary Media		111
	3.9.1	Formulation of the Boundary Value Problem	111
	3.9.2	Reduction of the Problem to an Integro-Functional Equation	112
	3.9.3	Solution of the Integro-Functional Equation	114
3.10	The Exactly Factorizable Linear Wave Equation with Memory and a Variable Coefficient		115
	3.10.1	Factorization Theorem	115
	3.10.2	Solution of the Boundary Value Problem	116

References 117

Subject Index 119

1. Nonlinear Waves in Homogeneous Media

In this chapter we present the prerequisite information about the solutions of nonlinear wave equations with constant coefficients. These equations describe the dynamics of longitudinal wave propagation in nonlinear elastic homogeneous rods. (Effects of transverse inertia and wave reflection from the lateral surface are neglected.) It should be noted that the dynamic equations obtained in this approximation are similar to the one-dimensional equations describing plane wave propagation in a nonbounded space. In this book, for ease of notation we deal with "waves in rods" rather than "plane waves in space". The main point of Chap. 1 is made in Theorem 1.3.2 about the exact factorization of the nonlinear wave equation with constant coefficients.

1.1 Preliminaries

1.1.1 Equations of Motion of a Homogeneous Nonlinear Rod

Let us consider a homogeneous rod of density ϱ in which the deformation ε and stress σ are related by the constitutive equation

$$\varepsilon = a(\sigma), \tag{1.1.1}$$

where $a(0) = 0$, $a'(\sigma) > 0$, $a''(\sigma) \neq 0$. We shall study longitudinal stress and deformation waves which propagate in such a rod after application of a load.

Suppose now that in some domain in which the time t and the Lagrangian coordinate x change, stress and deformation are smooth functions. Then we can write the equations of motion of the rod as follows:

$$\frac{1}{\varrho}\frac{\partial \sigma}{\partial x} = \frac{\partial v}{\partial t}, \quad \frac{\partial v}{\partial x} = \frac{\partial \varepsilon}{\partial t}.$$

Here v is the velocity of the material element of the rod. Substituting the value of ε from (1.1.1) into the latter equation we obtain a nonlinear system, which is well-known in mechanics

$$\frac{1}{\varrho}\frac{\partial \sigma}{\partial x} = \frac{\partial v}{\partial t}, \tag{1.1.2}$$

$$\frac{\partial v}{\partial x} = \frac{\partial a(\sigma)}{\partial t}. \tag{1.1.3}$$

(Similar systems also appear in gas dynamics and mechanics of fluids [1.1.2].)

It is obvious that the function v can be eliminated from our system of equations. To do this one should only differentiate (1.1.2) with respect to x and (1.1.3) with respect to t and add the results. Hence the system (1.1.2,3) will reduce to a single nonlinear wave equation

$$\frac{\partial^2 a(\sigma)}{\partial t^2} - \frac{1}{\varrho}\frac{\partial^2 \sigma}{\partial x^2} = 0. \tag{1.1.4}$$

Remark. If the functions $\sigma = \sigma(t,x)$, $\varepsilon = \varepsilon(t,x)$ are smooth in some domain of change of t and x, then the system (1.1.2,3) is equivalent to (1.1.4). But if the functions σ and ε are discontinuous, then (1.1.2,3) and (1.1.4) are slightly different due to the nonuniqueness of the solution of the wave equation (1.1.4). A more detailed discussion of this problem will be given below.

1.1.2 Riemann Invariants and Characteristics

The definition of the functions called the *Riemann invariants* of the system (1.1.2,3) is as follows:

$$r \equiv v + \int_0^\sigma \sqrt{a'(\sigma)/\varrho}\, d\sigma, \quad s \equiv v - \int_0^\sigma \sqrt{a'(\sigma)/\varrho}\, d\sigma. \tag{1.1.5}$$

One can easily show that the functions r and s satisfy the equations

$$\sqrt{\varrho a'(\sigma)}\,\frac{\partial r}{\partial t} - \frac{\partial r}{\partial x} = 0, \tag{1.1.6}$$

$$\sqrt{\varrho a'(\sigma)}\,\frac{\partial s}{\partial t} + \frac{\partial s}{\partial x} = 0. \tag{1.1.7}$$

Each of the equations (1.1.6,7) can be rewritten in an equivalent form using the *characteristics* of these equations. Let us consider for example (1.1.6). Temporarily treating the function $\sigma = \sigma(t,x)$ as a known quantity, in the t, x–plane we can write the following ordinary differential equation (equation of characteristics):

$$\frac{dt}{dx} = \frac{\sqrt{\varrho a'(\sigma)}}{-1}, \quad \sigma = \sigma(t,x).$$

The right-hand side of this equation is equal to the ratio of the coefficient to $\partial r/\partial t$ to the coefficient to $\partial r/\partial x$ taken from (1.1.6).

Then along each characteristic $t = t(x)$, defined by any solution of the previous equation, we obtain the following relation:

$$\frac{dr}{dx} = \frac{\partial r}{\partial t}\frac{dt}{dx} + \frac{\partial r}{\partial x} = \frac{\sqrt{\varrho a'(\sigma)}}{-1}\frac{\partial r}{\partial t} + \frac{\partial r}{\partial x} = 0.$$

Here we have used the rule for deriving the total derivative and have employed (1.1.6). The system

$$\frac{dt}{dx} = -\sqrt{\varrho a'(\sigma)}, \quad \frac{dr}{dx} = 0 \qquad (1.1.8)$$

represents the desired equivalent form of (1.1.6).

Analogously, (1.1.7) corresponds to the system

$$\frac{dt}{dx} = \sqrt{\varrho a'(\sigma)}, \quad \frac{ds}{dx} = 0. \qquad (1.1.9)$$

We also notice that by interchanging t and x, we obtain the following systems:

$$\frac{dx}{dt} = -\frac{1}{\sqrt{\varrho a'(\sigma)}}, \quad \frac{dr}{dt} = 0 \qquad (1.1.8')$$

and

$$\frac{dx}{dt} = \frac{1}{\sqrt{\varrho a'(\sigma)}}, \quad \frac{ds}{dt} = 0, \qquad (1.1.9')$$

which are equivalent to (1.1.8,9).]

It is obvious from (1.1.8) that along each characteristic of negative slope l^-, defined by the equation $dt/dx = -\left(\varrho a'(\sigma)\right)^{1/2}$, the value of r is constant. Analogously, from (1.1.9) it follows that the value of s is constant along each characteristic l^+ of positive slope, defined by the equation $dt/dx = \left(\varrho a'(\sigma)\right)^{1/2}$.

However, as long as σ is unknown, the families of characteristics l^- and l^+ are not yet defined. In spite of this the constancy of the Riemann invariants r and s along the corresponding curves is of great importance.

We also emphasize the fact that (1.1.8,9) for the invariants r and s hold only in the domain of smoothness of the functions v and σ. If the functions v and σ suffer jumps across some curve, then the invariants r and s will also suffer jumps across this curve.

1.1.3 Simple Wave Equation

Suppose that $v = v_0 = \text{const}$ for $x > 0$, $t = 0$ and let $v(t,x)$, $\sigma(t,x)$ be smooth functions of t, x. Then it is clear that in the domain covered by the family of characteristics of negative slope, issuing from the points of the semi-axis $x > 0$, we have

$$r \equiv \text{const} \qquad (1.1.10)$$

because r has one and the same value along each characteristic l^- belonging to the considered family (Fig. 1.1).

Now, according to (1.1.10), the first of the equalities (1.1.5) gives us the functional relation between v and σ

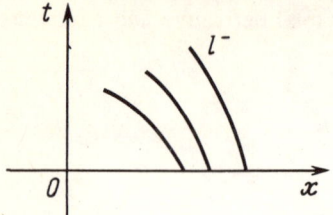

Fig. 1.1. The function $t(x)$ is given with l^- as a parameter

$$v = -\int_0^\sigma \sqrt{\frac{a'(\sigma)}{\varrho}}\, d\sigma + \text{const}\,. \tag{1.1.11}$$

Using this relation, we can eliminate v from the second equality in (1.1.5)

$$s = -1 \int_0^\sigma \sqrt{\frac{a'(\sigma)}{\varrho}}\, d\sigma + \text{const}\,.$$

Finally, substituting the obtained expression into (1.1.7), we get a first order differential equation for σ which holds in the domain where the Riemann invariant r is identically constant:

$$\sqrt{\varrho a'(\sigma)}\,\frac{\partial \sigma}{\partial t} + \frac{\partial \sigma}{\partial x} = 0\,. \tag{1.1.12}$$

Similarly, if $s \equiv \text{const}$ in some domain, then in such a domain the following equation for stress holds:

$$\sqrt{\varrho a'(\sigma)}\,\frac{\partial \sigma}{\partial t} - \frac{\partial \sigma}{\partial x} = 0\,. \tag{1.1.12'}$$

Equations (1.1.12) and (1.1.12') are called *equations of simple waves*. It is evident that (1.1.12) describes waves travelling to the right, and (1.1.12') describes waves travelling to the left.

1.1.4 Conditions on the Strong Shock

Suppose that across the curve $t = t(x)$ the functions ε, σ and v suffer finite jumps. Such discontinuities are called *strong shocks* (in contrast to weak shocks for which these functions are continuous while their derivatives are discontinuous). We shall call the curve $t = t(x)$ the *shock-wave front*. We introduce the following notation for the jump of the function $f(x)$ across the shock-wave front:

$$[f] \equiv f(t(x) + 0, x) - f(t(x) - 0, x)\,.$$

Let us denote the wave front velocity by U:

$$U \equiv \left(\frac{dt(x)}{dx}\right)^{-1}\,.$$

Then, as is well-known [1.3], the following relations for the jumps $[\varepsilon]$, $[\sigma]$ and $[v]$ must hold:

a) the *dynamic compatibility condition* which results from the law of conservation of momentum:

$$[\sigma] = -\varrho[v]U \ ; \qquad (1.1.13)$$

b) the *kinematic compatibility condition*:

$$[v] = -[\varepsilon]U \ ; \qquad (1.1.14)$$

c) the relation which directly follows from the constitutive equation (1.1.1):

$$[\varepsilon] = [a(\sigma)] \ . \qquad (1.1.15)$$

Eliminating $[v]$ and $[\varepsilon]$ from (1.1.13–15), we easily obtain for the front velocity the following condition expressed in terms of stress:

$$U = \sqrt{\frac{[\sigma]}{\varrho[a(\sigma)]}} \ . \qquad (1.1.16)$$

Here we have chosen the positive value of the square root; this choice corresponds to the shock-wave travelling to the right.

1.1.5 Stability Condition for the Strong Shock

Real strong shocks must satisfy the stability condition [1.4,5]

$$\frac{1}{\sqrt{\varrho a'(\sigma(t(x)-0,x))}} < U < \frac{1}{\sqrt{\varrho a'(\sigma(t(x)+0,x))}} \ . \qquad (1.1.17)$$

Hence, if the value $\left(\varrho a'(\sigma)\right)^{-1/2}$ is the local velocity of sound (in Lagrangian coordinates) then the velocity of the shock-wave U must be supersonic from the point of view of an observer standing in front of it, and subsonic for an observer behind it.

The condition (1.1.17) can be equivalently expressed as

$$\frac{1}{\sqrt{\varrho a'(\sigma(t(x), x+0))}} < U < \frac{1}{\sqrt{\varrho a'(\sigma(t(x), x-0))}} \ . \qquad (1.1.17')$$

Let us consider some qualitative aspects [1.6] which explain the meaning of this stability condition. Consider the profile of a stress wave containing a shock (Fig. 1.2a) and suppose, for a moment, that instead of the left-hand inequality (1.1.17′) the reverse inequality

$$\frac{1}{\sqrt{\varrho a'(\sigma(t(x), x+0))}} > U$$

holds.

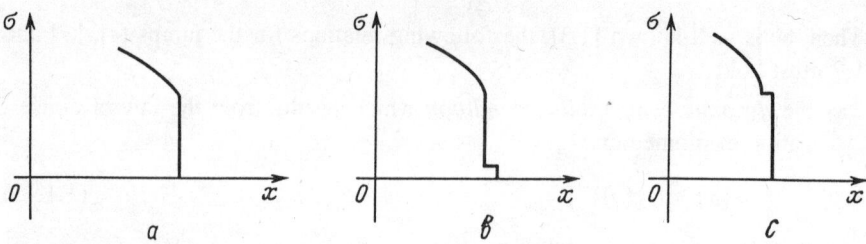

Fig. 1.2. For different cases $\sigma(x)$ is plotted, see **a** and **b** from *left* to *right*

However, according to (1.1.16) the small discontinuous profile disturbance represented in Fig. 1.2b, will propagate at a velocity close to

$$\left(\varrho a'\left(\sigma\left(t(x), x+0\right)\right)\right)^{-1/2},$$

and consequently will leave behind the main shock-wave front. A number of such small disturbances will destroy the shock-wave front.

Suppose now that instead of the right-hand inequality (1.1.17′) the reverse inequality

$$U > \frac{1}{\sqrt{\varrho a'(\sigma(t(x), x-0))}}$$

holds. However, according to (1.1.16) the small discontinuous disturbance, shown in Fig. 1.2c, will propagate at a velocity close to $\{\varrho a'\left(\sigma\left(t(x), x-0\right)\right)\}^{-1/2}$. According to the assumption made above this disturbance will fall behind the main shock-wave front. As in the previous case, a series of such disturbances will destroy the shock-wave front. Thus, the replacement of at least one of the inequalitites in (1.1.17′) [or, equivalently, in (1.1.17)] by the opposite one leads to the instability of the shock-wave.

In contrast, the validity of the stability condition (1.1.17′) [or (1.1.17)] leads to the following properties of the wave motion. In the situation shown in Fig. 1.2b the main shock-wave front catches up with the disturbance, and in the situation shown in Fig. 1.2c the disturbance catches up with the main shock-wave front. Consequently, the shock-wave turns out to be stable.

Remark. It is clear from (1.1.16,17) that the stability condition can also be rewritten as

$$[\sigma] > 0 \quad \text{if} \quad a''(\sigma) < 0,$$
$$[\sigma] < 0 \quad \text{if} \quad a''(\sigma) > 0. \tag{1.1.17″}$$

The case where $a''(\sigma)$ changes its sign is more complicated [1.7], and will not be discussed in these notes.

1.1.6 Weak Shocks

Suppose now that ε, σ and v are continuous in some domain, but their derivatives are discontinuous across the curve l, belonging to the interior of the domain. Thus, ε, σ and v suffer *weak shocks* on the curve l. A weak shock, however, can also be treated as a strong shock of infinitesimal amplitude. Going to the limit in (1.1.16) as $[\sigma] \to 0$, we deduce that the velocity of weak shock propagation equals $(\varrho a'(\sigma))^{-1/2}$, i.e., weak shocks propagate along the characteristics. It is remarkable that this fact can be derived directly from the system of equations (1.1.2,3) without using the conditions on the (strong) shock. For further details see [1.2].

1.2 Nonlinear Hyperbolic Equations of the First Order

1.2.1 Conditions on the Shock

Let us assume that $w = w(t, x)$ is a function which is smooth everywhere, except at a curve $t = t(x)$, across which it suffers a jump. We shall suppose that outside the curve $t = t(x)$ the nonlinear equation

$$\frac{\partial \varphi(w)}{\partial t} + \frac{\partial w}{\partial x} = 0 \tag{1.2.1}$$

holds, where $\varphi(0) = 0$, $\varphi'(w) > 0$, $\varphi''(w) \neq 0$. In addition, let us suppose that across the discontinuity curve $t = t(x)$ the condition

$$U = \frac{[w]}{[\varphi(w)]} \tag{1.2.2}$$

is valid, where $U \equiv (dt(x)/dx)^{-1}$.

We shall call (1.2.1,2) a *conservation law*. For the origin of conditions of the type of (1.2.2) for the laws of conservation of physical quantities, see [1.2,5].

It can be proved [1.2] that a discontinuous function w, satisfying (1.2.1,2), is a weak solution of (1.2.1) in the following sense: for an arbitrary smooth test function $g(t, x)$ with a compact support

$$-\iint [g'_t \varphi(w) + g'_x w] \, dt \, dx = 0 \, .$$

Conversely, a weak solution of (1.2.1), smooth outside the curve $t = t(x)$ and having finite one-sided limits as the point (t, x) tends to the curve, satisfies the condition (1.2.2) [and in the domain of smoothness is a classical solution of (1.2.1)].

As was done in Sect. 1.1.6, taking the limit of (1.2.2) as $[w] \to 0$, we find that weak shocks of the function w propagate along the characteristics $dt/dx = \varphi'(w)$ of (1.2.1).

1.2.2 Constancy of the Integrals of Solutions

The solutions of the conservation law (1.2.1,2) have the following important property: it turns out that if $w = 0$ as $t \to \pm\infty$, then

$$\int_{-\infty}^{\infty} w(t,x)\,dt = \text{const} \qquad (1.2.3)$$

holds.

In fact, if the family of the straight lines $x = \text{const}$ does not intersect the curve of discontinuity, then integrating (1.2.1) over t we immediately obtain

$$\varphi(w)\Big|_{t=-\infty}^{t=\infty} - \frac{d}{dx}\int_{-\infty}^{\infty} w\,dt = 0,$$

which proves the validity of (1.2.3) in this particular case.

Suppose now that the family of straight lines $x = \text{const}$ does intersect the curve of discontinuity. Then we reverse the reasoning given in [1.2]. By using (1.2.1) we obtain

$$\left(\int_{-\infty}^{t(x)-0} + \int_{t(x)+0}^{\infty}\right)\frac{\partial\varphi(w)}{\partial t}\,dt + \left(\int_{-\infty}^{t(x)-0} + \int_{t(x)+0}^{\infty}\right)\frac{\partial w}{\partial x}\,dt = 0. \qquad (1.2.4)$$

However, it is evident that

$$\left(\int_{-\infty}^{t(x)-0} + \int_{t(x)+0}^{\infty}\right)\frac{\partial\varphi(w)}{\partial t}\,dt = -[\varphi(w)] \qquad (1.2.5)$$

since $\varphi(w) = 0$ as $t \to \pm\infty$. On the other hand, it is clear that

$$\frac{d}{dx}\int_{-\infty}^{\infty} w\,dt = \frac{d}{dx}\left(\int_{-\infty}^{t(x)-0} + \int_{t(x)+0}^{\infty}\right) w\,dt$$

$$= w(t(x)-0, x)\frac{dt}{dx} + \int_{-\infty}^{t(x)-0}\frac{\partial w}{\partial x}\,dt$$

$$- w(t(x)+0, x)\frac{dt}{dx} + \int_{t(x)+0}^{\infty}\frac{\partial w}{\partial x}\,dt,$$

i.e.,

$$\left(\int_{-\infty}^{t(x)-0} + \int_{t(x)+0}^{\infty}\right)\frac{\partial w}{\partial x}\,dt = [w]\frac{dt(x)}{dx} + \frac{d}{dx}\int_{-\infty}^{\infty} w\,dt. \qquad (1.2.6)$$

Substituting now (1.2.5,6) into (1.2.4) we obtain

$$-[\varphi(w)] + [w]\frac{dt(x)}{dx} + \frac{d}{dx}\int_{-\infty}^{\infty} w\,dt = 0.$$

Hence, due to the condition on the shock (1.2.2), it follows that

$$\frac{d}{dx}\int_{-\infty}^{\infty} w\,dt = 0.$$

Thus, in this case (1.2.3) is also valid.

Finally, we should mention the case where the function $w(t,x)$ is smooth for example, for $0 < x < x_0$, and suffers a jump across the curve $t = t(x)$ for $x > x_0$. Then from the previous considerations it follows that

$$\int_{-\infty}^{\infty} w\,dt = C_1 \quad \text{for} \quad 0 < x < x_0,$$

$$\int_{-\infty}^{\infty} w\,dt = C_2 \quad \text{for} \quad x > x_0$$

where $C_1 = \text{const}$, $C_2 = \text{const}$. But it is evident that $w(t, x_0 - 0) = w(t, x_0 + 0)$, otherwise the straight line $x = x_0$ would be a discontinuity line across which the condition on the shock (1.2.2) is not valid. Therefore $C_1 = C_2$. Hence, (1.2.3) is established in the general case.

1.2.3 Solution of the Boundary Value Problem Method of Characteristics

Let us return to (1.2.1). At first let the function $w(t,x)$ be smooth. Then it is possible to rewrite (1.2.1) in the form

$$\varphi'(w)\frac{\partial w}{\partial t} + \frac{\partial w}{\partial x} = 0. \tag{1.2.7}$$

We shall consider the following boundary value problem for this equation:

$$w = 0 \quad \text{for} \quad x > 0, \quad t = 0,$$
$$w(t,0) = w_0(t), \tag{1.2.8}$$

where $w_0(t)$ is a smooth function defined on the semi-axis at $t \geq 0$ and equal to zero for $t > T > 0$. We shall also assume that the function $w_0(t)$ can be extended by zero to the semi-axis $t < 0$ without loss of smoothness; for the extended function we shall preserve the former notation $w_0(t)$.[1]

Let us show that in the domain of smoothness of the function w, the solution of the problem (1.2.7,8) can be analytically constructed by the method of characteristics.

Really, let us write the equations of characteristics for (1.2.7):

$$\frac{dt}{dx} = \varphi'(w), \quad \frac{dw}{dx} = 0. \tag{1.2.9}$$

---

[1] Such an extension of the boundary function is very convenient for the application of the method of characteristics. Later on, in problems of the type of (1.2.8) with zero initial data we shall suppose the boundary function to be extended by zero to the semi-axis $t < 0$

From the second equation of this system it is obvious that along each characteristic $w = $ const. Thus from the first equation of the system (1.2.9) it follows that

$$\frac{dt}{dx} = \text{const} .$$

Hence all the characteristics of (1.2.7) are straight lines.

The solution of the system (1.2.9), obviously, can be written in a parametric form:

$$w = w_0(\tau) ,$$
$$t = x\varphi'(w_0(\tau)) + \tau \qquad (1.2.10)$$

where τ is a parameter equal to the value of t at $x = 0$. Eliminating τ from (1.2.10), we can also get an implicit formula for w,

$$w = w_0\left(t - x\varphi'(w)\right) . \qquad (1.2.11)$$

Obviously, (1.2.10) defines $w = w(t,x)$. Let us now fix the variable x and consider w as a function of t. Clearly, one can consider the profile $w(t,x)$, $x = $ const, to be the result of evolution (with the growth of x) of the boundary profile $w_0(\tau)$. To understand this evolution better, let us fix some value $\tau > 0$ and the corresponding "height" of the profile $w_0(\tau)$. Then for $x = $ const > 0, as follows from (1.2.10), the same height of the profile will be achieved at $t = x\varphi'(w_0(\tau)) + \tau$. Thus the boundary profile $w_0(\tau)$ will become deformed, and so for sufficiently large values of x ($x > x_0$) the deformed profile $w(t,x)$, $x = $ const, will become multi-valued (such a process is called *overtaking* of the wave). In the domain $x > x_0$ we shall denote the solution of the system (1.2.10) [or, which is the same, for (1.2.11)] by $\tilde{w}(t,x)$.

Figure 1.3 shows an overtaking wave in the case where the boundary profile $w_0(\tau)$ is nonnegative and the growth of x implies the profile displacement (along t-axis) which grows with the height of the profile. Hence, the greater the boundary values are, the smaller their speeds.

Thus we can conclude that

a) the function $\varphi'(w)$ is increasing for $w \geq 0$ [see (1.2.9)];
b) along x-axis the direction of overtaking is opposite to the direction of the wave propagation.

If we consider the case where $\varphi''(w) < 0$, then the direction of overtaking along the x-axis would be opposite.

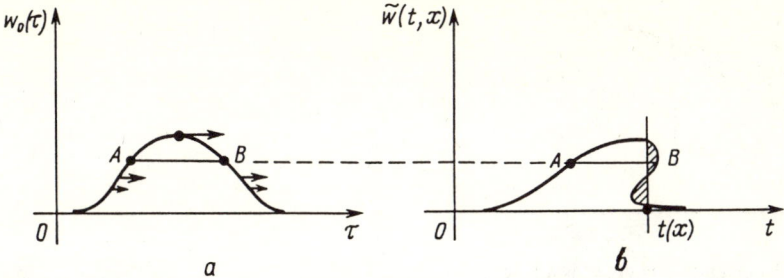

Fig. 1.3. The evolution of the profiles discussed in the text is given, see **a** and **b** from *left* to *right*

1.2.4 Wave Breaking

We now call the reader's attention to the following circumstance: the single-valued boundary profile $w_0(\tau)$ changed its form (with the growth of x) so that the boundary values travelled along characteristics of (1.2.7). Thus, the appearance of multivaluedness as a result of such a deformation of the profile means that characteristics of (1.2.7) intersect for x sufficiently large. It is clear that if $\tilde{w}(t, x)$ is a k-valued function in some domain, then k characteristics of (1.2.7) do intersect at each point of this domain. (For the boundary profile $w_0(\tau)$ having only one extremum, the deformed profile $\tilde{w}$ (Fig.1.3) is no more than three-valued. Hence in this case no more than three characteristics of the equation (1.2.7) may intersect for arbitrary t, x).

Furthermore, the origin of multivaluedness is clearly characterized by the fact that for some minimal $x = x_0 > 0$ there appears a vertical tangent to the profile. This effect is called *wave breaking*.

It is clear that wave breaking corresponds to the first intersection of characteristics (if one looks at it from the x-axis). It is also obvious that the first intersection will necessarily be the intersection of infinitely close characteristics. In fact, if the characteristics l_1 and l_2 starting from the points τ_1 and τ_2 of the t-axis intersect for some $x = x_1 > 0$, then the characteristic l starting from an arbitrary point $\tau \in (\tau_1, \tau_2)$ will necessarily intersect one of the characteristics l_1, l_2 for some $x = x'_1 < x_1$ (Fig.1.4). Hence the stated result follows.

Now it is clear that the point (t_0, x_0), at which the wave breaking occurs, belongs to the envelope of the family of characteristics of (1.2.7). It is also clear that in the t, x-plane for $x < x_0$ there are no points belonging to the envelope, i.e., x_0 is the minimal of the abscissas of the points of the envelope. The fact established is also valid in a more general case where characteristics of the equation are curvilinear and values of the solution of the equation are not constant along characteristics [1.2].

The abscissa of wave breaking equals [1.2]

$$x_0 = -\frac{1}{d/d\tau\ \varphi'(w_0(\tau))|_{\tau=\tau^*}} \tag{1.2.12}$$

where τ^* is a point at which the derivative $d\varphi'(w_0(\tau))/d\tau$ achieves its negative minimum; for simplicity we shall suppose this point to be a unique one. (If

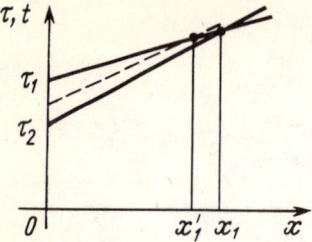

Fig. 1.4. The quantity τ is given as a function of x

$d\varphi'(w_0(\tau))/d\tau \geq 0$, then wave breaking does not occur at all.) The time of wave breaking can be calculated from (1.2.10) as follows:

$$t_0 = x_0\varphi'(w_0(\tau^*)) + \tau^* .$$

It can be shown that in the point (t_0, x_0) both derivatives w'_t and w'_x become infinite [1.2].

1.2.5 Principle of Equal Areas

For physical problems, as a rule, single-valued solutions are needed. The appearance of multivaluedness, which was described above, leads us to the conclusion that our assumption of smoothness of the solution is no longer acceptable, and the genuine solution of the problem has a shock [which must satisfy the already given condition (1.2.2)].

It turns out, however, that there exists a very simple method of constructing the discontinuous solution $w(t, x)$ of our problem (1.2.1,2,8) from the multi-valued function $\tilde{w}(t, x)$. This method is called the *principle of equal areas*.

Let us note, firstly, that according to (1.2.10) the transformation of the boundary profile $w_0(\tau)$ conserves the area bounded by the profile. In fact, this transformation conserves the lengths of the horizontal segments connecting the corresponding points of the profile. Furthermore, the distances from these horizontal segments to the time-axis also remain unchanged (Fig. 1.3). Hence our statement follows from the Fubini theorem.

Secondly, for the still unknown single-valued discontinuous solution, the following equality must hold [see (1.2.3)]:

$$\int_{-\infty}^{\infty} w \, dt = \text{const}$$

Now, it is not difficult to surmise that the genuine solution can be obtained from $\tilde{w}(t, x)$ by the following method: for each fixed $x > x_0$ a vertical line in $t, \tilde{w}$-plane must be drawn so that the domains shaded in Fig. 1.3b should have equal areas. (We shall denote the point of intersection of this vertical line with the t-axis by $t(x)$.) The resultant shock profile shown in Fig. 1.3b by a bold line corresponds to the desired discontinuous but single-valued solution of our problem.

The fact that the constructed function $w(t, x)$ satisfies the equation outside the line of discontinuity, $t = t(x)$, is evident. The validity of the initial and boundary conditions is also clear. Finally, it can be easily shown [1.2] that for the shock introduced above the condition on the shock (1.2.2) in fact holds and the velocity of this shock satisfies the following inequalities:

$$\frac{1}{\varphi'(w(t(x) - 0, x))} < U < \frac{1}{\varphi'(w(t(x) + 0, x))} \quad \text{with} \quad U \equiv \left(\frac{dt(x)}{dx}\right)^{-1}. \quad (1.2.13)$$

Remark 1. The arguments given in Sect. 1.1.5 show us that (1.2.13) is the stability condition for the shock-wave front. It is clear from (1.2.2) that (1.2.13) can be rewritten in the following equivalent form:

$$[w] > 0 \quad \text{if} \quad \varphi''(w) < 0,$$
$$[w] < 0 \quad \text{if} \quad \varphi''(w) > 0. \quad (1.2.13')$$

Remark 2. If the boundary function $w_0(\tau)$ has the form shown in Fig. 1.3a (i.e., it vanishes outside a segment and has a unique extremum), then it is geometrically evident that the function $t = t(x)$, $x > x_0$ describing the propagation of the front will be continuous, single-valued and increasing. It is also evident that the value of the shock $[w]$ will continuously vary along the front, but will not go to zero in a finite time. It can be shown that $[w] \sim \text{const} \times t^{-1/2}$, $t \to \infty$ [1.2,8,9].

1.2.6 An Example

The example given below is well-known [1.2]. Let $\varphi(w) = w + kw^2/2$, $k > 0$, and let

$$w = 0 \quad \text{for} \quad x > 0, \ t = 0,$$
$$w(t, 0) = \Theta(t) \quad \text{where} \quad \Theta(t) = 1 \quad \text{for} \quad t \geq 0,$$
$$= 0 \quad \text{for} \quad t < 0.$$

Then the solution of (1.2.1), given by the method of characteristics, is continuous for $x > 0$ and has the form

$$w_1(t, x) = 1 \quad \text{for} \quad t > (1 + k)x,$$
$$w_1(t, x) = \frac{t/x - 1}{k} \quad \text{for} \quad x < t < (1 + k)x,$$
$$w_1(t, x) = 0 \quad \text{for} \quad t < x.$$

On the other hand, under the same initial and boundary conditions the equation (1.2.1) has another solution which contains a shock

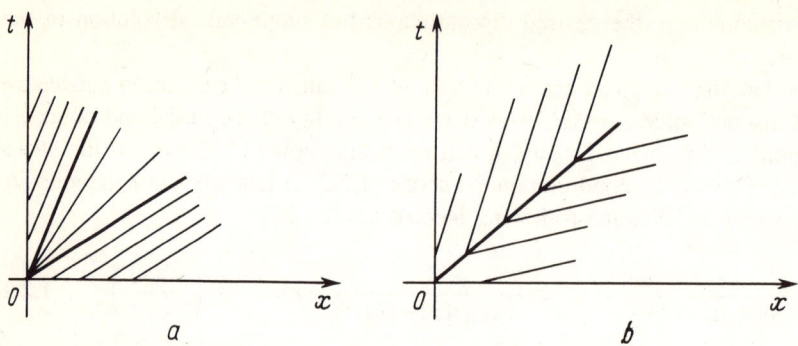

Fig. 1.5. For different situations $t(x)$ is plotted, see **a** and **b** from *left* to *right*

$$w_2(t,x) = 1 \quad \text{for} \quad t > \left(1 + \frac{k}{2}\right) x\,,$$

$$w_2(t,x) = 0 \quad \text{for} \quad t < \left(1 + \frac{k}{2}\right) x\,.$$

It is evident that the solution w_2 satisfies the condition on the shock

$$U = \frac{[w]}{[\varphi(w)]} = \frac{1}{1 + k/2}\,.$$

It is clear, however, that $w_2(t,x)$ does not satisfy the stability condition (1.2.13). Considering a continuous boundary profile close to $\Theta(t)$, we can see that the corresponding unique solution of the problem (constructed by the method of characteristics) will be close to $w_1(t,x)$ and hence will essentially differ from the unstable discontinuous solution $w_2(t,x)$.

Thus, in our example it is natural to consider $w_1(t,x)$ as the genuine solution. The characteristics of w_1 and w_2 are drawn in Figs. 1.5a and 1.5b, respectively.

1.2.7 Ordinary Differential Equation for a Shock Propagating into an Undisturbed Domain

Now, let us consider the case where $w = 0$ in the domain which lies in front of the wave front. (It is clear that in this case the shock should be given directly in the boundary condition.) Then the differential equation for the line of the wave front can be easily obtained by the following reasoning which does not use the principle of equal areas.

From the implicit relation (1.2.11), which takes the form

$$\tilde{w} = w_0(t - x\varphi'(\tilde{w}))$$

in the domain of multivaluedness, we conclude that, in the case considered, for the genuine discontinuous solution w the following relation is valid on the front:

$$[w] = w_0(t - x\varphi'([w])) \,. \tag{1.2.14}$$

Joining to the above relation the condition on the shock (1.2.2) which in the case considered has the form

$$\frac{dt(x)}{dx} = \frac{\varphi([w])}{[w]} \,, \tag{1.2.15}$$

we obtain a system for two unknown functions, $t(x)$ and $[w]$.

One can easily eliminate the shock $[w]$ from the system (1.2.14, 15). In fact, let us solve (1.2.14) for $[w]$ choosing the root of maximal module

$$[w] = G(t, x) \,.$$

Then, by (1.2.15), it follows that

$$\frac{dt}{dx} = \frac{\varphi(G(t, x))}{G(t, x)} \,. \tag{1.2.16}$$

If the shock-wave front issues out of the point $x = 0$ at the moment $t = 0$, then (1.2.16) should be supplemented with the evident condition

$$t(0) = 0 \,. \tag{1.2.17}$$

The solution $t(x)$ of the problem (1.2.16,17) can be constructed by Euler's method with an arbitrary degree of accuracy. Then the value of $[w]$ can be obtained from the equation (1.2.14).

Let us note that the method of describing the shock-wave front given above may be applied to more general cases where the principle of equal areas does not hold.

1.3 Exact Factorization of the Nonlinear Wave Equation with Constant Coefficients

1.3.1 Introductory Observations

Turning to stress waves in the nonlinear elastic rod, we return to the nonlinear wave equation for stress (Sect. 1.1):

$$\frac{\partial^2 a(\sigma)}{\partial t^2} - \frac{1}{\varrho}\frac{\partial^2 \sigma}{\partial x^2} = 0 \tag{1.3.1}$$

which holds in the domain of smoothness of $\sigma(t, x)$.

On the other hand, we have seen that if for the system of equations (1.1.2,3) the Riemann invariant r is identically constant in some domain, then in this domain the stress σ satisfies the equation of simple waves travelling to the right, i.e.,

$$\sqrt{a'(\sigma)}\,\frac{\partial\sigma}{\partial t} + \frac{1}{\sqrt{\varrho}}\,\frac{\partial\sigma}{\partial x} = 0 . \tag{1.3.2}$$

We have also seen that if in some domain the Riemann invariant s is identically constant, then in this domain σ satisfies the equation of simple waves travelling to the left, i.e.,

$$\sqrt{a'(\sigma)}\,\frac{\partial\sigma}{\partial t} - \frac{1}{\sqrt{\varrho}}\,\frac{\partial\sigma}{\partial x} = 0 . \tag{1.3.3}$$

Since the nonlinear wave equation (1.3.1) always holds [in the domain of smoothness of $\sigma(t,x)$], then each smooth solution of (1.3.2,3) will obviously be the solution of (1.3.1). (In the domain of smoothness of the function $\sigma(t,x)$ the equations (1.3.2,3) can be analytically solved by the method of characteristics, as was shown in Sect. 1.2.3). Therefore we conclude that the first order differential operators from the left-hand sides of (1.3.2) and (1.3.3) may be singled out as multipliers in the nonlinear wave operator (1.3.1).

1.3.2 Factorization Theorem for the Wave Equation for Stress

Theorem 1.3.2 [1.10]. The equation (1.3.1) can be factorized in one of the two following ways:

$$\left\{\frac{\partial}{\partial t}\sqrt{a'(\sigma)} \mp \frac{1}{\sqrt{\varrho}}\frac{\partial}{\partial x}\right\}\left\{\sqrt{a'(\sigma)}\frac{\partial}{\partial t} \pm \frac{1}{\sqrt{\varrho}}\frac{\partial}{\partial x}\right\}\sigma = 0 \tag{1.3.4}$$

(simultaneously we choose either upper or lower signs).

Remark. All quantities in braces on the left-hand side of (1.3.4) are considered to be operators (of differentiation or multiplication by some function). It is also supposed that operators in their products act in turn from the right to the left. For example,

$$\frac{\partial}{\partial t}\sqrt{a'(\sigma)}\sqrt{a'(\sigma)}\frac{\partial}{\partial t}\sigma \equiv \frac{\partial}{\partial t}\left(a'(\sigma)\frac{\partial\sigma}{\partial t}\right) .$$

Proof. By directly expanding the left-hand side of (1.3.4), we obtain

$$\left\{\frac{\partial}{\partial t}\sqrt{a'(\sigma)} \mp \frac{1}{\sqrt{\varrho}}\frac{\partial}{\partial x}\right\}\left\{\sqrt{a'(\sigma)}\frac{\partial\sigma}{\partial t} \pm \frac{1}{\sqrt{\varrho}}\frac{\partial\sigma}{\partial x}\right\}$$

$$= \frac{\partial}{\partial t}\sqrt{a'(\sigma)}\sqrt{a'(\sigma)}\frac{\partial\sigma}{\partial t} \pm \frac{1}{\sqrt{\varrho}}\frac{\partial}{\partial t}\sqrt{a'(\sigma)}\,\frac{\partial\sigma}{\partial x}$$

$$\mp \frac{1}{\sqrt{\varrho}}\frac{\partial}{\partial x}\sqrt{a'(\sigma)}\,\frac{\partial\sigma}{\partial t} - \frac{1}{\varrho}\frac{\partial^2\sigma}{\partial x^2} . \tag{1.3.5}$$

In accordance with the usual convention on the order of acting of operators, the first term on the right-hand side of (1.3.5) equals

1.3 Exact Factorization of the Nonlinear Wave Equation with Constant Coefficients

$$\frac{\partial}{\partial t} a'(\sigma) \frac{\partial \sigma}{\partial t} = \frac{\partial^2 a(\sigma)}{\partial t^2} \, . \tag{1.3.6}$$

Hence, to prove the theorem we only have to establish that

$$\frac{1}{\sqrt{\varrho}} \frac{\partial}{\partial t} \sqrt{a'(\sigma)} \, \frac{\partial \sigma}{\partial x} = \frac{1}{\sqrt{\varrho}} \frac{\partial}{\partial x} \sqrt{a'(\sigma)} \, \frac{\partial \sigma}{\partial t} \, . \tag{1.3.7}$$

However, the equality (1.3.7) clearly follows from the fact that

$$\frac{1}{\sqrt{\varrho}} \frac{\partial}{\partial t} \frac{\partial}{\partial x} \int_0^\sigma \sqrt{a'(\sigma)} \, d\sigma = \frac{1}{\sqrt{\varrho}} \frac{\partial}{\partial x} \frac{\partial}{\partial t} \int_0^\sigma \sqrt{a'(\sigma)} \, d\sigma \, .$$

Thus, from (1.3.5–7) it follows that

$$\left\{ \frac{\partial}{\partial t} \sqrt{a'(\sigma)} \mp \frac{1}{\sqrt{\varrho}} \frac{\partial}{\partial x} \right\} \left\{ \sqrt{a'(\sigma)} \frac{\partial \sigma}{\partial t} \pm \frac{1}{\sqrt{\varrho}} \frac{\partial \sigma}{\partial x} \right\}$$
$$\equiv \frac{\partial^2 a(\sigma)}{\partial t^2} - \frac{1}{\varrho} \frac{\partial^2 \sigma}{\partial x^2} \, ,$$

which completes the proof.

1.3.3 Difference Between Linear and Nonlinear Factorization

To be precise, let us consider the case where in the factorization (1.3.4) the upper signs are taken:

$$\left\{ \frac{\partial}{\partial t} \sqrt{a'(\sigma)} - \frac{1}{\sqrt{\varrho}} \frac{\partial}{\partial x} \right\} \left\{ \sqrt{a'(\sigma)} \frac{\partial \sigma}{\partial t} + \frac{1}{\sqrt{\varrho}} \frac{\partial \sigma}{\partial x} \right\} = 0 \, . \tag{1.3.8}$$

Unfortunately, such a representation of (1.3.1) does not enable us to construct the general solution of (1.3.1). In fact, setting

$$w = \sqrt{a'(\sigma)} \frac{\partial \sigma}{\partial t} + \frac{1}{\sqrt{\varrho}} \frac{\partial \sigma}{\partial x} \tag{1.3.9}$$

we should then solve the equation corresponding to the external multiplier in (1.3.8) as follows:

$$\frac{\partial}{\partial t} \left(\sqrt{a'(\sigma)} w \right) - \frac{1}{\sqrt{\varrho}} \frac{\partial w}{\partial x} = 0 \, . \tag{1.3.10}$$

However, the coefficients of this equation still depend on σ which is unknown to us. (Here the basic difference of the factorization is compared with the factorization of linear equations.) Thus, to construct the general solution of (1.3.8), we should solve both (1.3.9) and (1.3.10) simultaneously, which is as difficult to do as solving the original system of dynamic equations (1.1.2,3). Later on, however, we shall see that the factorization of the nonlinear wave equation is not useless.

1.3.4 Factorization Theorem for the Deformation Wave Equation

Let us resolve the constitutive equation $\varepsilon = a(\sigma)$ with respect to σ by setting

$$\sigma = b(\varepsilon) . \tag{1.3.11}$$

Then from (1.3.1) we immediately obtain the nonlinear wave equation for deformation, i.e.,

$$\frac{\partial^2 \varepsilon}{\partial t^2} - \frac{1}{\varrho} \frac{\partial^2 b(\varepsilon)}{\partial x^2} = 0 . \tag{1.3.12}$$

This equation evidently belongs to the same type of equations as (1.3.1); the only difference between (1.3.1) and (1.3.12) is that t and x play opposite roles. Hence the following result is valid.

Theorem 1.3.4. The equation (1.3.12) can be factorized in one of the two following ways:

$$\left\{ \frac{\partial}{\partial t} \mp \frac{1}{\sqrt{\varrho}} \frac{\partial}{\partial x} \sqrt{b'(\varepsilon)} \right\} \left\{ \frac{\partial}{\partial t} \pm \frac{1}{\sqrt{\varrho}} \sqrt{b'(\varepsilon)} \frac{\partial}{\partial x} \right\} \varepsilon = 0 . \tag{1.3.13}$$

Clearly, the substitution $\varepsilon = a(\sigma)$ [where $a(\sigma) \equiv b^{-1}(\sigma)$] converts the factorization (1.3.13) into (1.3.4).

1.3.5 Earnshaw's Theorem

Finally, from the equation of motion and the relation (1.3.11) it is not difficult to obtain the following equation for displacement:

$$\frac{\partial^2 u}{\partial t^2} - \frac{1}{\varrho} \frac{\partial}{\partial x} b\left(\frac{\partial u}{\partial x} \right) = 0 . \tag{1.3.14}$$

Theorem 1.3.5. (Earnshaw [1.11]). Let

$$g(y) \equiv \frac{1}{\sqrt{\varrho}} \int_0^y \sqrt{b'(y)} \, dy + \text{const} . \tag{1.3.15}$$

Then each smooth solution of any of the two following equations

$$\frac{\partial u}{\partial t} + g\left(\frac{\partial u}{\partial x} \right) = 0 , \tag{1.3.16}$$

$$\frac{\partial u}{\partial t} - g\left(\frac{\partial u}{\partial x} \right) = 0 \tag{1.3.17}$$

also satisfies (1.3.14).

1.3 Exact Factorization of the Nonlinear Wave Equation with Constant Coefficients

Proof. Let, for example, $u(t,x)$ be a solution of (1.3.17). Then let us demonstrate that $u(t,x)$ also satisfies (1.3.14). Substituting $\partial u/\partial t$ from (1.3.17) into (1.3.14) and applying (1.3.15), we obtain:

$$A \equiv \frac{\partial}{\partial t} g\left(\frac{\partial u}{\partial x}\right) - \frac{1}{\varrho}\frac{\partial}{\partial x} b\left(\frac{\partial u}{\partial x}\right) = g'\left(\frac{\partial u}{\partial x}\right)\frac{\partial^2 u}{\partial x \partial t} - \frac{1}{\varrho}b'\left(\frac{\partial u}{\partial x}\right)\frac{\partial^2 u}{\partial x^2}$$

$$= g'\left(\frac{\partial u}{\partial x}\right)\frac{\partial}{\partial x} g\left(\frac{\partial u}{\partial x}\right) - \frac{1}{\varrho}b'\left(\frac{\partial u}{\partial x}\right)\frac{\partial^2 u}{\partial x^2}$$

$$= \left\{\left(g'\left(\frac{\partial u}{\partial x}\right)\right)^2 - \frac{1}{\varrho}b'\left(\frac{\partial u}{\partial x}\right)\right\}\frac{\partial^2 u}{\partial x^2} = 0.$$

Hence, the stated result follows.

Remark. One can easily see that (1.3.16,17) describe simple waves in terms of displacement.

1.3.6 Generalization of Earnshaw's Theorem

Let a rod be subjected to the uniformly distributed longitudinal external load $F(t)$ counted at a unit of length. Then, the equation for displacement will take the form

$$\frac{\partial^2 u}{\partial t^2} - \frac{1}{\varrho}\frac{\partial}{\partial x} b\left(\frac{\partial u}{\partial x}\right) = \frac{F(t)}{\varrho}. \tag{1.3.18}$$

The following result generalizing Theorem 1.3.5 is true:

Theorem 1.3.6. Each smooth solution of any of the two following equations

$$\frac{\partial u}{\partial t} + g\left(\frac{\partial u}{\partial x}\right) = \frac{1}{\varrho}\int_0^t F(t)dt, \tag{1.3.19}$$

$$\frac{\partial u}{\partial t} - g\left(\frac{\partial u}{\partial x}\right) = \frac{1}{\varrho}\int_0^t F(t)dt \tag{1.3.20}$$

also satisfies (1.3.18). Here the function g is defined by (1.3.15).

Proof. Let, for example, $u(t,x)$ satisfy (1.3.20). Let us check that in this case $u(t,x)$ also satisfies (1.3.18). Substituting $\partial u/\partial t$ from (1.3.20) into the left-hand side of (1.3.18), we obtain consecutively

$$\frac{\partial}{\partial t}\left\{g\left(\frac{\partial u}{\partial x}\right) + \frac{1}{\varrho}\int_0^t F(t)dt\right\} - \frac{1}{\varrho}\frac{\partial}{\partial x}\,b\left(\frac{\partial u}{\partial x}\right)$$

$$= g'\left(\frac{\partial u}{\partial x}\right)\frac{\partial}{\partial x}\frac{\partial u}{\partial t} + \frac{F(t)}{\varrho} - \frac{1}{\varrho}b'\left(\frac{\partial u}{\partial x}\right)\frac{\partial^2 u}{\partial x^2}$$

$$= g'\left(\frac{\partial u}{\partial x}\right)\frac{\partial}{\partial x}\left\{g\left(\frac{\partial u}{\partial x}\right) + \frac{1}{\varrho}\int_0^t F(t)dt\right\} + \frac{F(t)}{\varrho} - \frac{1}{\varrho}b'\left(\frac{\partial u}{\partial x}\right)\frac{\partial^2 u}{\partial x^2}$$

$$= \left(g'^2 - \frac{b'}{\varrho}\right)\frac{\partial^2 u}{\partial x^2} + \frac{F(t)}{\varrho} \equiv \frac{F(t)}{\varrho}.$$

Thus we have proved that the function $u(t,x)$, satisfying (1.3.20), also satisfies (1.3.18). Analogously, we can prove the assertion of the theorem about (1.3.19). The theorem is proved.

1.3.7 A Boundary Value Problem Posed in Terms of Displacements

Now let the function $F(t)$ be identically equal to zero for $t \leq 0$ and let us pose the following problem for (1.3.18):

$$u = \frac{\partial u}{\partial t} = 0 \quad \text{for} \quad x > 0, \quad t = 0,$$
$$u(t,0) = u_0(t). \tag{1.3.21}$$

Here $u_0(t)$ is supposed to be smooth and identically equal to zero for $t \leq 0$.

From the physical point of view, it is clear that before wave breaking occurs, the solution of our problem will have a single-wave character. Hence, we have the equation (1.3.19) describing waves travelling to the right. Note now that from the first of the conditions (1.3.21) it follows that

$$\frac{\partial u}{\partial x} = \frac{\partial u}{\partial t} = 0 \quad \text{for} \quad x > 0, \quad t = 0.$$

Therefore, putting $t = 0$ in (1.3.19), we easily obtain that const $= 0$ in (1.3.15). Thus,

$$g(y) \equiv \frac{1}{\sqrt{\varrho}}\int_0^y \sqrt{b'(y)}\,dy. \tag{1.3.22}$$

Since (1.3.19) is not quasi-linear, it is impossible to solve (1.3.19,21) directly by the method of characteristics. However, we shall see that the method of characteristics can be applied to this problem if we reformulate it in terms of deformation.

Note first of all that by putting $x = 0$ in (1.3.19), we immediately obtain the boundary condition for deformation:

$$u_0'(t) + g(\varepsilon_0(t)) = \frac{1}{\varrho}\int_0^t F(t)dt \tag{1.3.23}$$

where

1.3 Exact Factorization of the Nonlinear Wave Equation with Constant Coefficients

$\varepsilon_0(t) \equiv \partial u/\partial x|_{t=0}$.

The equality (1.3.23) can also be rewritten in the following equivalent form:

$$\varepsilon_0(t) = g^{-1}\left(\frac{1}{\varrho}\int_0^t F(t)dt - u_0'(t)\right) . \tag{1.3.24}$$

Obviously, in accordance with the assumptions given above, (1.3.24) is identically equal to zero for $t \leq 0$.

Furthermore, from the first of the relations (1.3.21) it is clear that

$$\varepsilon = \partial\varepsilon/\partial t = 0 \quad \text{for} \quad x > 0, \ t = 0. \tag{1.3.25}$$

The equalities (1.3.24,25) are consequently the boundary and initial conditions for our problem, formulated in terms of deformation.

Finally, the nonlinear single-wave equation describing the propagation of deformation can be obtained by differentiating (1.3.19) with respect to x:

$$\frac{\partial\varepsilon}{\partial t} + \sqrt{\frac{b'(\varepsilon)}{\varrho}}\frac{\partial\varepsilon}{\partial x} = 0 . \tag{1.3.26}$$

The problem (1.3.24–26) can be easily integrated by the method of characteristics given in Sect. 1.2.3. In fact, the equations of characteristics for (1.3.26) have the form

$$\frac{dt}{dx} = \sqrt{\frac{\varrho}{b'(\varepsilon)}} , \qquad \frac{d\varepsilon}{dx} = 0 .$$

Therefore, using the boundary condition for deformation we obtain

$$\varepsilon(t,x) = \varepsilon_0(\tau) ,$$

$$t = x\sqrt{\frac{\varrho}{b'(\varepsilon_0(\tau))}} + \tau . \tag{1.3.27}$$

(Since $\varepsilon_0(t) = 0$ for $t \leq 0$, the initial conditions for deformation will hold automatically.)

Now, because of the boundary condition in (1.3.21), we finally obtain the solution of the problem posed above in terms of displacement

$$u(t,x) = \int_0^x \varepsilon(t,x)dx + u_0(t) \tag{1.3.28}$$

where the function $\varepsilon(t,x)$ is parametrically given by the relations (1.3.27).

Let us now check that (1.3.28) really is the desired solution of our problem. In fact, the validity of (1.3.21) is evident (since $u_0(0) = u_0'(0) = 0$ by the assumptions made above). Let us show that (1.3.28) satisfies the equation of motion (1.3.18). By Theorem 1.3.6 it is sufficient to demonstrate that this function satisfies the single-wave equation for displacement (1.3.19). By differentiating (1.3.28) with respect to t, we obtain

$$\frac{\partial u}{\partial t} = u_0'(t) + \int_0^x \frac{\partial \varepsilon}{\partial t} dx .$$

However, from (1.3.26) it follows that

$$\frac{\partial \varepsilon}{\partial t} = -\sqrt{\frac{b'(\varepsilon)}{\varrho}} \frac{\partial \varepsilon}{\partial x} ;$$

hence

$$\frac{\partial u}{\partial t} = u_0'(t) - \frac{1}{\sqrt{\varrho}} \int_0^x \sqrt{b'(\varepsilon)} \frac{\partial \varepsilon}{\partial x} dx = u_0'(t) - g(\varepsilon(t,x)) + g(\varepsilon_0(t)) ,$$

i.e.,

$$\frac{\partial u}{\partial t} + g\left(\frac{\partial u}{\partial x}\right) = u_0'(t) + g(\varepsilon_0(t)) = \frac{1}{\varrho} \int_0^t F(t) dt ,$$

which gives the result required.

Note that the constructed solution is valid only until the wave breaking occurs.

1.4 Shock-Wave in a Simple System

1.4.1 Formulation of the Problem

As we know from Sect. 1.1, outside the shock-wave front the stress satisfies the following equation in a homogeneous nonlinear rod of density ϱ:

$$\frac{\partial^2 a(\sigma)}{\partial t^2} - \frac{1}{\varrho} \frac{\partial^2 \sigma}{\partial x^2} = 0 , \qquad (1.4.1)$$

and on the shock (which should be travelling to the right) the following condition must hold:

$$U = \sqrt{\frac{[\sigma]}{\varrho[a(\sigma)]}} . \qquad (1.4.2)$$

In Sect. 1.4. we shall impose the following restrictions on the function $a(\sigma)$:

$$a(0) = 0, \quad a'(\sigma) > 0, \quad a''(\sigma) < 0 .$$

Let the rod be semi-infinite and located on the semi-axis $x \geq 0$; now consider the following model problem:

$$\begin{aligned} \sigma = \partial\sigma/\partial t = 0 \quad &\text{for} \quad x > 0, \quad t = 0 , \\ \sigma(t,0) = \sigma_0(t) . & \end{aligned} \qquad (1.4.3)$$

Here it is supposed that

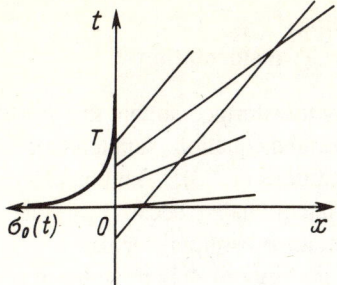

Fig. 1.6. The functions $t(x)$ and $\sigma_0(t)$ are shown

$\sigma_0(t) = 0$ for $t < 0$ and for $t > T > 0$,

$\sigma_0(0) > 0$,

$\sigma_0'(t) \leq 0$ for $t > 0$,

i.e., a tension wave is considered and the boundary profile has a "triangular" shape (Fig. 1.6).

Note now that if we are able to solve the equation

$$\sqrt{a'(\sigma)}\,\frac{\partial \sigma}{\partial t} + \frac{1}{\sqrt{\varrho}}\frac{\partial \sigma}{\partial x} = 0 \tag{1.4.4}$$

under the conditions (1.4.2,3), then, in accordance with Theorem 1.3.2, we should obtain the desired solution of (1.4.1–3) formulated in terms of stress. (Below, we shall see that such a single-wave solution is not quite adequate in terms of the physical sense of the problem, and should be considered only as a good approximation to the genuine solution, valid for moderate amplitudes.)

In Fig. 1.6 the characteristics of (1.4.4) which correspond to the boundary profile $\sigma_0(t)$ are represented. It is evident from Fig. 1.6 that the greater the values of σ, the greater their speeds. Therefore the tension wave will steepen in the direction of its propagation along the x-axis (Sect. 1.2.3).

1.4.2 Nonconformity of the Single-Wave Equation to the Shock Condition

It is rather clear that the solution of (1.4.4) satisfying the condition on the shock (1.4.2) will not be a generalized solution of (1.4.4). In fact, the generalized solution should satisfy the following condition on the shock (Sect. 1.2.1):

$$U = \frac{[\sigma]/\sqrt{\varrho}}{\left[\int_0^\sigma \sqrt{a'(\sigma)}d\sigma\right]}$$

which evidently never coincides with (1.4.2) (except in the case where $a(\sigma)$ is a linear function.)

Thus (1.4.2,4) is not a conservation law and the solution of the problem (1.4.2–4) cannot be directly obtained by the method of equal areas.

1.4.3 Transformation of the Single-Wave Equation. Integral Equation for $g(\sigma)$ Generating the Transformation

It turns out, however, that (1.4.4) can be identically transformed so that the principle of equal areas will be applicable to the transformed equation [supplemented with the condition on the shock (1.4.2) and the conditions (1.4.3)]. The boundary and initial conditions (1.4.3) are mentioned here intentionally, because the fact that the shock propagates into the undisturbed domain is essential for us.

Let the function $\sigma(t,x)$ be smooth. Then we multiply (1.4.4) by some unknown nonnegative function $g(\sigma)$. This leads us to the equality

$$g(\sigma)\sqrt{a'(\sigma)}\,\frac{\partial \sigma}{\partial t} + \frac{1}{\sqrt{\varrho}}\,g(\sigma)\,\frac{\partial \sigma}{\partial x} = 0$$

whence

$$\frac{\partial}{\partial t}\int_0^\sigma g(\sigma)\sqrt{a'(\sigma)}\,d\sigma + \frac{1}{\sqrt{\varrho}}\frac{\partial}{\partial x}\int_0^\sigma g(\sigma)d\sigma = 0 \ . \tag{1.4.5}$$

Our purpose is to find such a function $g(\sigma)$ that in the problem considered the condition on the shock (1.4.2) would coincide with the condition

$$U = \frac{\left[\int_0^\sigma g(\sigma)d\sigma\right]/\sqrt{\varrho}}{\left[\int_0^\sigma g(\sigma)\sqrt{a'(\sigma)}d\sigma\right]} \ . \tag{1.4.6}$$

Due to the assumption made above, the shock wave will propagate into the undisturbed domain; then in our problem the condition of coincidence of (1.4.2) with (1.4.6) will take the following form:

$$\sqrt{\frac{[\sigma]}{\varrho a([\sigma])}} = \frac{\int_0^{[\sigma]} g(\sigma)d\sigma/\sqrt{\varrho}}{\int_0^{[\sigma]} g(\sigma)\sqrt{a'(\sigma)}d\sigma} \ . \tag{1.4.7}$$

[Note that in the general case, where the shock wave propagates in the predisturbed domain, it is impossible to obtain the coincidence of (1.4.2) and (1.4.6)].

If the function $g(\sigma)$ satisfying (1.4.7) is found, then, on the one hand, it is evident that the solution of (1.4.3,5,6) will also be the solution of (1.4.1–3). On the other hand, it is possible to find the solution of (1.4.3,5,6) by the method of equal areas, reducing this to the problem considered in Sect. 1.2.5.

In fact, let

$$G(\sigma) \equiv \int_0^\sigma g(\sigma)d\sigma \ , \quad F(\sigma) \equiv \int_0^\sigma g(\sigma)\sqrt{a'(\sigma)}d\sigma$$

and let us introduce a new unknown function

$$w = G(\sigma) \ .$$

Then (1.4.5) will take the form

$$\frac{\partial}{\partial t} F(G^{-1}(w)) + \frac{1}{\sqrt{\varrho}} \frac{\partial w}{\partial x} = 0 \,; \qquad (1.4.8)$$

consequently, the condition (1.4.6) will become

$$U = \frac{[w]/\sqrt{\varrho}}{[F(G^{-1}(w))]} \,. \qquad (1.4.9)$$

The initial and boundary conditions for w can be found from the corresponding conditions for σ (1.4.3):

$$w = 0 \quad \text{for} \quad x > 0, \quad t = 0,$$

$$w(t,0) = G(\sigma_0(t)) \,. \qquad (1.4.10)$$

The problem (1.4.8–10) clearly belongs to the type considered in Sect. 1.2.5 and the solution may be constructed with the help of the principle of equal areas.

As a result, when the solution of (1.4.8–10) is found, it remains to set

$$\sigma = G^{-1}(w) \,. \qquad (1.4.11)$$

One can easily demonstrate that the solution constructed will also satisfy the stability condition (1.1.17).

1.4.4 Construction of the Function $g(\sigma)$

Thus, it remains to find the function g which satisfies the integral equation (1.4.7). For conciseness we shall write σ instead of $[\sigma]$ in further calculations. Then, (1.4.7) will take the form

$$\int_0^\sigma g(\sigma)\sqrt{a'(\sigma)}\,d\sigma = \sqrt{\frac{a(\sigma)}{\sigma}} \int_0^\sigma g(\sigma)\,d\sigma \,.$$

By differentiating, we obtain

$$g(\sigma)\sqrt{a'(\sigma)} = \left(\sqrt{\frac{a(\sigma)}{\sigma}}\right)' \int_0^\sigma g(\sigma)\,d\sigma + \sqrt{\frac{a(\sigma)}{\sigma}}\,g(\sigma) \,,$$

i.e.,

$$\frac{g(\sigma)}{\int_0^\sigma g(\sigma)\,d\sigma} = \frac{\left(\sqrt{a(\sigma)/\sigma}\right)'}{\sqrt{a'(\sigma)} - \sqrt{a(\sigma)/\sigma}}$$

or,

$$\frac{g(\sigma)}{\int_0^\sigma g(\sigma)\,d\sigma} = \frac{1}{2\sigma}\left(1 + \sqrt{\frac{\sigma a'(\sigma)}{a(\sigma)}}\right) \,. \qquad (1.4.12)$$

Using again the notation

$$G(\sigma) \equiv \int_0^\sigma g(\sigma)d\sigma,$$

we obtain from (1.4.12) that

$$G(\sigma) = \exp \int_{\sigma^*}^\sigma \frac{1}{2\sigma} \left(1 + \sqrt{\frac{\sigma a'(\sigma)}{a(\sigma)}}\right) d\sigma \qquad (1.4.13)$$

where $\sigma^* = \text{const} > 0$.

The fact that (1.4.13) really can be represented in a form of the integral

$$\int_0^\sigma g(\sigma)d\sigma,$$

where $g(\sigma)$ is some nonnegative function, is a consequence of the following

Lemma 1.4.4. Let $a(\sigma)$ be a smooth function such that $a(0) = 0$ and $a'(\sigma) > 0$, $a''(\sigma) < 0$ for $\sigma \geq 0$. Then for the function $G(\sigma)$ defined in (1.4.13) the following propositions hold:

a) $G(0) = 0$;
b) $G(\sigma)$ is monotone increasing with σ for $\sigma \geq 0$;
c) $G'(\sigma)$ is a locally bounded function for $\sigma \geq 0$.

Proof. For small $\sigma > 0$ we have

$$\left(\frac{\sigma a'(\sigma)}{a(\sigma)}\right)^{1/2} = \left(\frac{\sigma(a'(0) + \sigma a''(0) + \ldots)}{\sigma a'(0) + \frac{\sigma^2}{2} a''(0) + \ldots}\right)^{1/2}$$

$$= \left(\frac{a'(0) + \mathcal{O}(\sigma)}{a'(0) + \mathcal{O}(\sigma)}\right)^{1/2} = 1 + \mathcal{O}(\sigma).$$

Hence for small $\sigma > 0$

$$\frac{1}{2\sigma}\left(1 + \sqrt{\frac{\sigma a'(\sigma)}{a(\sigma)}}\right) = \frac{1}{\sigma} + \mathcal{O}(1). \qquad (1.4.14)$$

The expression on the left-hand side of (1.4.14) is evidently bounded if $\sigma > 0$ is bounded away from zero, thus we shall consider the equality (1.4.14) to be performed for all $\sigma > 0$ (and not only for small $\sigma > 0$), meaning by O(1) some bounded in σ function.

From (1.4.14) it follows that

$$\int_{\sigma^*}^\sigma \frac{1}{2\sigma}\left(1 + \sqrt{\frac{\sigma a'(\sigma)}{a(\sigma)}}\right) d\sigma = \ln\frac{\sigma}{\sigma^*} + \int_{\sigma^*}^\sigma \mathcal{O}(1)d\sigma.$$

Substituting the above expression into (1.4.13), we obtain

$$G(\sigma) = \frac{\sigma}{\sigma^*} \exp \int_{\sigma^*}^{\sigma} \mathcal{O}(1) d\sigma , \qquad (1.4.15)$$

which completes the proof of the statement (a) of the lemma.

Furthermore, to prove the statement (b), it suffices to note that the integrand in (1.4.13) is positive.

Finally, as one can easily see, it suffices to prove the statement (c) for small $\sigma > 0$. Differentiating (1.4.13) and taking into account (1.4.14,15), we obtain

$$G'(\sigma) = \frac{1}{2\sigma} \left(1 + \sqrt{\frac{\sigma a'(\sigma)}{a(\sigma)}} \right) G(\sigma)$$

$$= \left(\frac{1}{\sigma} + \mathcal{O}(1) \right) \frac{\sigma}{\sigma^*} \exp \int_{\sigma^*}^{\sigma} \mathcal{O}(1) d\sigma , \quad \sigma \to +0 ,$$

which yields the local boundedness of $G'(\sigma)$. The lemma is proved.

Thus we have really found the nonnegative function $g(\sigma)$ satisfying (1.4.7):

$$g(\sigma) = \frac{dG(\sigma)}{d\sigma} = \frac{1}{2\sigma} \left(1 + \sqrt{\frac{\sigma a'(\sigma)}{a(\sigma)}} \right)$$

$$\times \exp \int_{\sigma^*}^{\sigma} \frac{1}{2\sigma} \left(1 + \sqrt{\frac{\sigma a'(\sigma)}{a(\sigma)}} \right) d\sigma , \qquad (1.4.16)$$

which completes the constructions of Sect. 1.4.3.

1.4.5 Discussion of the Results

We have constructed the analytical solution $\sigma = G^{-1}(w)$ of problem (1.4.1–3), formulated exclusively in terms of stress. In doing this we took into account Theorem 1.3.2 which enables us to replace the nonlinear wave equation (1.4.1) by its factor, i.e., the simple wave equation

$$\sqrt{a'(\sigma)} \frac{\partial \sigma}{\partial t} + \frac{1}{\sqrt{\varrho}} \frac{\partial \sigma}{\partial x} = 0 . \qquad (1.4.17)$$

Let us see, however, whether the constructed solution σ of problem (1.4.1–3) is in accordance with the solution of the original system of dynamic equations

$$\frac{\partial \sigma}{\partial x} = \varrho \frac{\partial v}{\partial t} , \qquad (1.4.18)$$

$$\frac{\partial v}{\partial x} = \frac{\partial a(\sigma)}{\partial t} \qquad (1.4.19)$$

and the corresponding conditions on the shock:

$$[\sigma] = -\varrho[v]U , \qquad (1.4.20)$$

$$[v] = -[a(\sigma)]U \qquad (1.4.21)$$

(Sect. 1.1.4).

First of all, it follows from (1.4.17,18) that behind the line of the front

$$\frac{\partial}{\partial t}\int_0^\sigma \sqrt{a'(\sigma)}d\sigma + \sqrt{\varrho}\frac{\partial v}{\partial t} = 0,$$

i.e.,

$$\frac{\partial r}{\partial t} = 0 \qquad (1.4.22)$$

[where r is the Riemann invariant; see (1.1.5)]. However, we know that in the complement to the front the invariant r satisfies the equation

$$\sqrt{\varrho a'(\sigma)}\frac{\partial r}{\partial t} - \frac{\partial r}{\partial x} = 0. \qquad (1.4.23)$$

From (1.4.22,23) it evidently follows that behind the front

$$\frac{\partial r}{\partial t} = 0, \quad \frac{\partial r}{\partial x} = 0$$

whence $r = $ const. Hence, behind the front the following relation must hold:

$$v = \text{const} - \int_0^\sigma \sqrt{\frac{a'(\sigma)}{\varrho}}\, d\sigma.$$

Since in front of the wave front $v = \sigma = 0$, we get from the previous relation that on the wave front we must have

$$[v] = \text{const} - \int_0^{[\sigma]} \sqrt{\frac{a'(\sigma)}{\varrho}}\, d\sigma. \qquad (1.4.24)$$

On the other hand, eliminating U from (1.4.20,21) and taking into account the fact that in front of the wave front $v = \sigma = 0$, we evidently obtain

$$[v] = -\sqrt{\frac{[\sigma]a([\sigma])}{\varrho}}. \qquad (1.4.25)$$

Now from (1.4.24,25) it follows that

$$\text{const} - \int_0^{[\sigma]} \sqrt{\frac{a'(\sigma)}{\varrho}}\, d\sigma = -\sqrt{\frac{[\sigma]a([\sigma])}{\varrho}}.$$

Setting in the previous relation $[\sigma] = 0$, we immediately obtain const = 0. Hence, in particular it follows that behind the front we must have

$$v = -\int_0^\sigma \sqrt{\frac{a'(\sigma)}{\varrho}}\, d\sigma.$$

1.4 Shock-Wave in a Simple System

Thus, we have to investigate whether the equality

$$\int_0^{[\sigma]} \sqrt{a'(\sigma)} d\sigma - \sqrt{[\sigma]a([\sigma])} = 0 \tag{1.4.26}$$

is possible. It is not difficult to verify that (1.4.26) cannot hold for any function except the linear one. Hence, the pair of functions

$$\sigma = G^{-1}(w), \quad v = -\int_0^\sigma \sqrt{\frac{a'(\sigma)}{\varrho}} d\sigma \tag{1.4.27}$$

(for the function $a(\sigma)$ different from a linear one) does not satisfy the conditions on the shock (1.4.20,21).

Note, however, that if the function $a(\sigma)$ can be represented in the form

$$a(\sigma) = a_1\sigma + a_2\sigma^2 + \ldots ; \quad a_1 > 0$$

then the difference on the left-hand side of (1.4.26) satisfies the following estimate:

$$\int_0^{[\sigma]} \sqrt{a'(\sigma)} d\sigma - \sqrt{[\sigma]a([\sigma])} = \mathcal{O}\left([\sigma]^3\right) . \tag{1.4.28}$$

This means that the pair of functions (1.4.27) satisfies the condition on the shock (1.4.20,21) with acciracy to $\mathcal{O}\left([\sigma]^3\right)$.

Finally, it is evident that (1.4.27) satisfies the equations of motion. Thus, for moderate amplitudes (1.4.27) gives a good approximation to the solution of (1.4.18–21) with initial and boundary conditions (1.4.3).

Let us also note the following important fact. Under the assumptions made above, the problem (1.4.18–21) has the unique solution $\sigma = \sigma^*(t, x)$, $v = v^*(t, x)$ satisfying the stability condition (1.1.17) [1.13].

It is clear that the function $\sigma = \sigma^*(t, x)$ [as well as the function $\sigma = G^{-1}(w)$] will be the solution of (1.4.1–3), but for the nonlinear function $a(\sigma)$

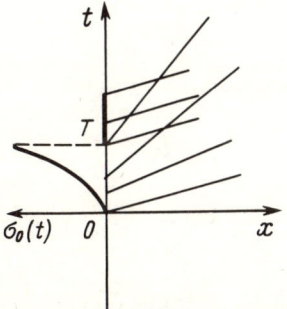

Fig. 1.7. The functions $t(x)$ and $\sigma(t)$ are illustrated

$$\sigma^*(t,x) \neq G^{-1}(w) \, .$$

Thus, we have found that the problem (1.4.1–3), posed exclusively in terms of stress, has more than one solution even under the additional stability condition for the strong shock (1.1.17). Also, one can easily see that the genuine solution $\sigma = \sigma^*(t,x)$ does not satisfy the simple wave equation, because the unique solution of (1.4.2–4) is evidently $G^{-1}(w)$. However, the above-mentioned means that a genuine shock-wave front gives rise to a comparatively weak wave travelling to the left. Clearly, we could not discover it by replacing (1.4.1) by its factor (1.4.4).

Remark. Likewise, we could consider the case where $a''(\sigma) > 0$, but in doing so we should take the boundary profile represented in Fig. 1.7. In this figure there are also characteristics of (1.4.4), corresponding to the given boundary profile. One can see in Fig. 1.7 that the greater values of σ are, the smaller their speeds; hence the tension wave travelling to the right along the rod will steepen backward. Therefore, the undisturbed domain will be left behind the wave front (contrary to the case considered above). Evidently, all the constructions of Sect. 1.4. may be carried over to this case. The compression wave (for which $\sigma \leq 0$) with boundary profile of triangular shape can be studied in the same way.

1.5 The Shock-Wave in a Simple System (Continuation)

1.5.1 Application of the Principle of Equal Areas

Formula (1.4.28) means that the approximate solution (1.4.27) differs from the exact one by the quantity $\mathcal{O}([\sigma]^3)$. Therefore, for example, it follows from (1.4.21) that the value of the front velocity U is defined with accuracy of order $\mathcal{O}([\sigma]^3)$. However, if we are limited by the accuracy of order $\mathcal{O}(\sigma^2)$, then it is not even necessary to suppose that the front propagates into the undisturbed domain, and the problem set in Sect. 1.4 can be solved in a more simple way. (As usual, we suppose that $a(0) = 0$, $a'(\sigma) > 0$, $a''(\sigma) \neq 0$.)

Indeed [1.2], it is not difficult to see that if

$$a(\sigma) = a_1\sigma + a_2\sigma^2 + \dots ,$$

then

$$\sqrt{\frac{[\sigma]}{[a(\sigma)]}} - \frac{[\sigma]}{[\int_0^\sigma \sqrt{a'(\sigma)}d\sigma]} = \mathcal{O}(\sigma^2) \, . \tag{1.5.1}$$

Hence, the condition on the shock (1.4.2) can be replaced with the mentioned accuracy by the condition

$$U = \frac{[\sigma]/\sqrt{\varrho}}{\left[\int_0^\sigma \sqrt{a'(\sigma)}d\sigma\right]} \cdot \tag{1.5.2}$$

The simple wave equation

$$\sqrt{a'(\sigma)}\frac{\partial\sigma}{\partial t} + \frac{1}{\sqrt{\varrho}}\frac{\partial\sigma}{\partial x} = 0, \tag{1.5.3}$$

supplemented with (1.5.2), represents a conservation law (Sect. 1.2.1), and the equation of the wave front can be obtained by the method of equal areas.

1.5.2 Application of Euler's Method

The method given in Sect. 1.5.1 is useful because it allows us to describe the shock-wave front propagation in a predisturbed domain. (Notice that namely this situation immediately follows the breaking of an originally smooth wave.) However, this method limits the possibilities of the method of introducing the shock into the simple wave.

Let us consider one more way of describing the shock-wave front (as above, we shall use the simple wave approximation). Let t_0, x_0 be the coordinates of the wave breaking. Then the value $\sigma(t_0, x_0)$ can still be found by the method of characteristics, and the jump $[\sigma(t_0, x_0)]$ is equal to zero. Hence, for the point (t_0, x_0), the condition on the shock gives

$$U = U_0 = \lim_{[\sigma]\to 0} \sqrt{\frac{[\sigma]}{\varrho[a(\sigma)]}} = \frac{1}{\sqrt{\varrho a'(\sigma(t_0, x_0))}} \cdot$$

Let

$$t_1 = t_0 + \Delta, \quad \text{and} \quad x_1 = x_0 + U_0\Delta$$

hold, where Δ is sufficiently small and positive (Fig. 1.8a). Now, let us draw the multivalued profile $\tilde{\sigma} = \tilde{\sigma}(t, x_1)$, given parametrically by the equations of characteristics, and put

$$U_1 = \sqrt{\frac{\sigma_1^+ - \sigma_1^-}{\varrho(a(\sigma_1^+) - a(\sigma_1^-))}}$$

(the quantities σ_1^+ and σ_1^- are determined in the way shown in Fig. 1.8b). Then we obtain the coordinates of the endpoint of the next segment of the broken line as

$$t_2 = t_1 + \Delta, \quad \text{and} \quad x_2 = x_1 + U_1\Delta$$

etc. The desired approximation to the line of the wave front will be a broken line composed of segments with endpoints at $(t_0, x_0), (t_1, x_1), \ldots$.

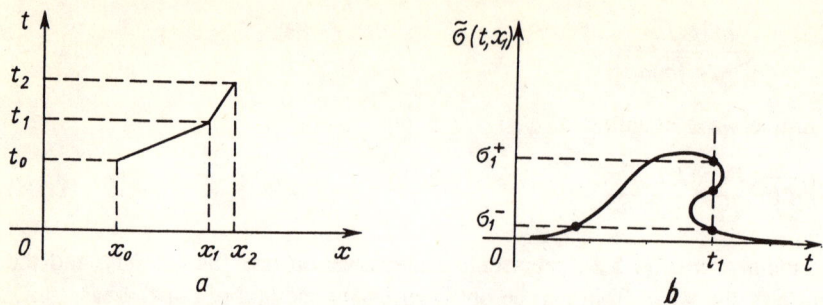

Fig. 1.8. Illustration to the application of Euler's method, see **a** and **b** from *left* to *right*

2. Nonlinear Short Waves of Finite Amplitude in Inhomogeneous Media

This chapter is devoted to factorization theorems for nonlinear wave equations with variable coefficients. The obtained mathematical results are applied to the investigation of short waves in inhomogeneous rods.

2.1 Asymptotic Factorization of the Nonlinear Wave Equation with a Variable Coefficient

In Sect. 2.1 we give a very important representation for the nonlinear wave equation describing stress propagation in a one-dimensional medium with variable density. In the case where the duration of the boundary pulse is very short, this representation gives the uniform asymptotic factorization of the wave equation under consideration. We shall also see that a single-wave equation, appearing as a result of such factorization, can be analytically integrated.

2.1.1 Representation of the Nonlinear Wave Equation with a Variable Coefficient

Let us consider a nonlinear elastic rod in which deformation ε and stress σ are related by the constitutive equation

$$\varepsilon = a(\sigma) , \qquad (2.1.1)$$

where $a'(\sigma) > 0$, $a(0) = 0$. As far as the rod density is concerned, we suppose that

$$\varrho = \varrho(x) > 0 .$$

The characteristic scale of inhomogeneity of the density is supposed to be equal to L. Then the equations of motion of the rod in the Lagrange coordinates will have the form

$$\frac{\partial v}{\partial t} = \frac{1}{\varrho(x)} \frac{\partial \sigma}{\partial x} ,$$

$$\frac{\partial v}{\partial x} = \frac{\partial a(\sigma)}{\partial t} , \qquad (2.1.2)$$

where v is the velocity of a material element. From (2.1.2) we easily obtain the desired nonlinear wave equation describing stress propagation in a rod of variable density:

$$\frac{\partial^2 a(\sigma)}{\partial t^2} - \frac{\partial}{\partial x} \frac{1}{\varrho(x)} \frac{\partial \sigma}{\partial x} = 0. \tag{2.1.3}$$

By analogy with Theorem 1.3.2 we shall now try to factorize (2.1.3) up to terms which do not contain derivatives of the unknown function σ.

Namely, we shall try to find the representation of the operator from (2.1.3) in the form

$$\begin{aligned} 0 &= \frac{\partial^2 a(\sigma)}{\partial t^2} - \frac{\partial}{\partial x} \frac{1}{\varrho(x)} \frac{\partial \sigma}{\partial x} \\ &\equiv \left\{ \frac{\partial}{\partial t} \sqrt{a'(\sigma)} - \frac{\partial}{\partial x} \frac{1}{\sqrt{\varrho(x)}} + g(x)\varphi'(\sigma) \right\} \\ &\quad \times \left\{ \sqrt{a'(\sigma)} \frac{\partial \sigma}{\partial t} + \frac{1}{\sqrt{\varrho(x)}} \frac{\partial \sigma}{\partial x} + g(x)\varphi(\sigma) \right\} - N(x, \sigma), \end{aligned} \tag{2.1.4}$$

where $g(x)$, $\varphi(\sigma)$ and $N(x, \sigma)$ are functions which are to be determined. For simplicity we restrict ourselves to the case where the inner multiplier in the product of braces describes waves travelling to the right. Multiplying out the braces on the right-hand side of (2.1.4), we can rewrite the right-hand side of (2.1.4) in the form

$$\begin{aligned} &\frac{\partial^2 a(\sigma)}{\partial t^2} - \frac{\partial}{\partial x} \frac{1}{\sqrt{\varrho(x)}} \sqrt{a'(\sigma)} \frac{\partial \sigma}{\partial t} + g(x)\varphi'(\sigma)\sqrt{a'(\sigma)} \frac{\partial \sigma}{\partial t} \\ &+ \frac{\partial}{\partial t} \sqrt{a'(\sigma)} \frac{1}{\sqrt{\varrho(x)}} \frac{\partial \sigma}{\partial x} - \frac{\partial}{\partial x} \frac{1}{\sqrt{\varrho(x)}} \frac{1}{\sqrt{\varrho(x)}} \frac{\partial \sigma}{\partial x} \\ &+ g(x)\varphi'(\sigma) \frac{1}{\sqrt{\varrho(x)}} \frac{\partial \sigma}{\partial x} + \frac{\partial}{\partial t} \sqrt{a'(\sigma)}\, g(x)\varphi(\sigma) \\ &- \frac{\partial}{\partial x} \frac{1}{\sqrt{\varrho(x)}} g(x)\varphi(\sigma) + g^2(x)\varphi'(\sigma)\varphi(\sigma) - N(x, \sigma). \end{aligned}$$

By combining the second addend with the fourth one, the third with the seventh, and the sixth with the eighth, we can rewrite the previous expression as

$$\frac{\partial^2 a(\sigma)}{\partial t^2} - \frac{\partial}{\partial x}\frac{1}{\varrho(x)}\frac{\partial \sigma}{\partial x}$$
$$+ \left\{ g(x)\varphi'(\sigma)\sqrt{a'(\sigma)}\,\frac{\partial \sigma}{\partial t} + \frac{\partial}{\partial t}\sqrt{a'(\sigma)}\,g(x)\varphi(\sigma) \right\}$$
$$- \left\{ \frac{\partial}{\partial x}\frac{1}{\sqrt{\varrho(x)}}\sqrt{a'(\sigma)}\frac{\partial \sigma}{\partial t} - \frac{\partial}{\partial t}\sqrt{a'(\sigma)}\frac{1}{\sqrt{\varrho(x)}}\frac{\partial \sigma}{\partial x} \right\}$$
$$+ \left\{ g(x)\varphi'(\sigma)\frac{1}{\sqrt{\varrho(x)}}\frac{\partial \sigma}{\partial x} - \frac{\partial}{\partial x}\frac{1}{\sqrt{\varrho(x)}}g(x)\varphi(\sigma) \right\}$$
$$+ g^2(x)\varphi'(\sigma)\varphi(\sigma) - N(x,\sigma)$$
$$\equiv \frac{\partial^2 a(\sigma)}{\partial t^2} - \frac{\partial}{\partial x}\frac{1}{\varrho(x)}\frac{\partial \sigma}{\partial x}$$
$$+ g(x)\left\{ \varphi'(\sigma)\sqrt{a'(\sigma)} + \frac{d}{d\sigma}\left(\sqrt{a'(\sigma)}\varphi(\sigma)\right) \right\}\frac{\partial \sigma}{\partial t}$$
$$- \left(\frac{1}{\sqrt{\varrho(x)}}\right)'_x \sqrt{a'(\sigma)}\frac{\partial \sigma}{\partial t} - \left(\frac{g(x)}{\sqrt{\varrho(x)}}\right)'_x \varphi(\sigma)$$
$$+ g^2(x)\varphi'(\sigma)\varphi(\sigma) - N(x,\sigma) \,. \tag{2.1.5}$$

Clearly, it is necessary now to get rid of those addends on the right-hand side of (2.1.5) which contain the derivative $\partial \sigma/\partial t$.

To do this, let us set

a) $\quad g(x) = \dfrac{d}{dx}\left(\dfrac{1}{\sqrt{\varrho(x)}}\right) \,,$ \hfill (2.1.6)

b) $\quad \varphi'(\sigma)\sqrt{a'(\sigma)} + \dfrac{d}{d\sigma}\left(\sqrt{a'(\sigma)}\varphi(\sigma)\right) = \sqrt{a'(\sigma)} \,.$ \hfill (2.1.7)

The equality (2.1.7) yields

$$2\varphi'(\sigma)\sqrt{a'(\sigma)} + \left(\sqrt{a'(\sigma)}\right)' \varphi(\sigma) = \sqrt{a'(\sigma)} \,,$$

i.e.,

$$\varphi' + \frac{1}{4}\frac{a''(\sigma)}{a'(\sigma)}\varphi = \frac{1}{2} \,.$$

The solution of this equation, vanishing for $\sigma = 0$, has the form

$$\varphi(\sigma) = \frac{1}{2}\int_0^\sigma \left(\frac{a'(\eta)}{a'(\sigma)}\right)^{1/4} d\eta \,. \tag{2.1.8}$$

After the functions $g(x)$ and $\varphi(\sigma)$ are defined, it remains to set

$$N(x,\sigma) = -\left(\frac{g(x)}{\sqrt{\varrho(x)}}\right)'_x \varphi(\sigma) + g^2(x)\varphi'(\sigma)\varphi(\sigma)$$

or, by applying (2.1.6,8), we obtain

$$N(x,\sigma) = \frac{1}{4\varrho^3(x)} \int_0^\sigma \left(\frac{a'(\eta)}{a'(\sigma)}\right)^{1/4} d\eta$$
$$\times \left\{ \varrho''\varrho - 2(\varrho')^2 + \frac{1}{4}(\varrho')^2 \frac{d}{d\sigma} \int_0^\sigma \left(\frac{a'(\eta)}{a'(\sigma)}\right)^{1/4} d\eta \right\} . \qquad (2.1.9)$$

Now, for the functions $g(x)$, $\varphi(\sigma)$, $N(x,\sigma)$ obtained above, the right-hand side of (2.1.5) takes exactly the form of the nonlinear wave operator from (2.1.3), which gives the representation required.

In the same manner, we can obtain such a representation of the operator from (2.1.3) that the inner multiplier in the product of braces will describe waves travelling to the left:

$$\frac{\partial^2 a(\sigma)}{\partial t^2} - \frac{\partial}{\partial x} \frac{1}{\varrho(x)} \frac{\partial \sigma}{\partial x}$$
$$\equiv \left\{ \frac{\partial}{\partial t} \sqrt{a'(\sigma)} + \frac{\partial}{\partial x} \frac{1}{\sqrt{\varrho(x)}} - g(x)\varphi'(\sigma) \right\}$$
$$\times \left\{ \sqrt{a'(\sigma)} \frac{\partial \sigma}{\partial t} - \frac{1}{\sqrt{\varrho(x)}} \frac{\partial \sigma}{\partial x} - g(x)\varphi(\sigma) \right\} - N(x,\sigma) ,$$

where functions $g(x)$, $\varphi(\sigma)$, $N(x,\sigma)$ are still defined by (2.1.6, 8, 9).

The representation of the nonlinear wave operator which was obtained above yields the following:

Theorem 2.1.1 [2.1]. The equation (2.1.3) can be represented in the form

$$\left\{ \frac{\partial}{\partial t} \sqrt{a'(\sigma)} \mp \left(\frac{\partial}{\partial x} \frac{1}{\sqrt{\varrho(x)}} + \frac{1}{4} \frac{\varrho'}{\varrho^{3/2}} \frac{d}{d\sigma} \int_0^\sigma \left(\frac{a'(\eta)}{a'(\sigma)}\right)^{1/4} d\eta \right) \right\}$$
$$\times \left\{ \sqrt{a'(\sigma)} \frac{\partial \sigma}{\partial t} \pm \left(\frac{1}{\sqrt{\varrho(\sigma)}} \frac{\partial \sigma}{\partial x} - \frac{1}{4} \frac{\varrho'}{\varrho^{3/2}} \int_0^\sigma \left(\frac{a'(\eta)}{a'(\sigma)}\right)^{1/4} d\eta \right) \right\}$$
$$= N(x,\sigma) , \qquad (2.1.10)$$

where the difference $N(x,\sigma)$ is given by (2.1.9). In the decomposition (2.1.10) either upper or lower signs are simultaneously chosen.

Remark. The requirement $\varphi(0) = 0$ is imposed in (2.1.8) for the discrepancy of factorization $N(x,\sigma)$ to vanish for $\sigma = 0$. Then, in case of a pulse propagating in the undisturbed medium, the discrepancy of factorization will be distinct from zero only in the domain where the wave amplitude is nonzero. This case will be considered below.

If we considered the case where the stress pulse propagates in a homogeneously prestressed medium, then instead of (2.1.8) we should, evidently, set

$$\varphi(\sigma) = \frac{1}{2} \int_{\bar{\sigma}}^{\sigma} \left(\frac{a'(\eta)}{a'(\sigma)} \right)^{1/4} d\eta ,$$

where $\bar{\sigma}$ = const is the initial value of stress.

2.1.2 Formulation of the Boundary Value Problem. Conditions of Asymptotic Factorization

Let us now set the following problem for the equation (2.1.3):

$$\sigma = \frac{\partial \sigma}{\partial t} = 0 \quad \text{for} \quad t = 0, \quad x > 0,$$

$$\sigma(t, 0) = \sigma_0 \left(\frac{t}{\gamma} \right). \tag{2.1.11}$$

Here $\sigma_0(\tau)$ is a smooth function equal to zero for $\tau < 0$ and for $\tau > T > 0$; γ is a small parameter

$$0 < \gamma \ll 1. \tag{2.1.12}$$

Thus the duration of the stress pulse appearing in our problem equals γT by the order of value.

Now, let us try to estimate the ratio of the discrepancy $N(x, \sigma)$ to the terms of the left-hand side of the nonlinear wave equation (2.1.3) for the short stress pulse under consideration. Consider, for example, the first addend on the left-hand side of (2.1.3). Since the pulse duration equals γT by the order of value, then

$$\frac{\partial^2 a(\sigma)}{\partial t^2} \sim \frac{a'(0)\sigma}{(\gamma T)^2} .$$

The same order of value has the second addend on the left-hand side of (2.1.3). However, as is clear from (2.1.9),

$$N(x, \sigma) \sim \frac{\sigma}{L^2 \varrho_c} ,$$

where L is the characteristic scale of inhomogeneity, ϱ_c is the characteristic value of density.

Hence, the ratio $N(x, \sigma)$ to the terms on the left-hand side of (2.1.3) is of order $O(\gamma)^2$ for the short pulse under consideration.

Thus, when solving (2.1.11) about the propagation of a short stress pulse, we can neglect the discrepancy $N(x, \sigma)$ in the representation (2.1.10) of the nonlinear wave equation.

Since our problem means the consideration of a wave travelling to the right, it is now clear that the wave equation (2.1.3) can be replaced by the first order equation which corresponds to the inner multiplier on the left-hand side of (2.1.10) when one is choosing the upper signs

$$\sqrt{a'(\sigma)}\frac{\partial\sigma}{\partial t} + \frac{1}{\sqrt{\varrho(x)}}\frac{\partial\sigma}{\partial x} - \frac{1}{4}\frac{\varrho'}{\varrho^{3/2}}\int_0^\sigma \left(\frac{a'(\eta)}{a'(\sigma)}\right)^{1/4} d\eta = 0 . \qquad (2.1.13)$$

In fact, by force of Theorem 2.1.1 each solution of this equation will also satisfy (2.1.3) with accuracy to the discrepancy $N(x, \sigma)$. Moreover, as we shall see below, it is always possible to construct a solution of (2.1.13) satisfying the conditions (2.1.11).

2.1.3 Single-Wave Solution of the Boundary Value Problem

We shall solve (2.1.11,13) without assuming the amplitudes to be small. Therefore, our analysis is possible only up to the moment of wave breaking (since, as we know, for large amplitudes the emergence of a shock corresponds to the origin of a two-wave process).

Let us write the equations of characteristics for (2.1.13) as follows:

$$\frac{dt}{dx} = \sqrt{\varrho(x)a'(\sigma)} , \qquad (2.1.14)$$

$$\frac{d\sigma}{dx} = \frac{1}{4}\frac{\varrho'(x)}{\varrho(x)}\int_0^\sigma \left(\frac{a'(\eta)}{a'(\sigma)}\right)^{1/4} d\eta . \qquad (2.1.15)$$

From (2.1.15) it follows that along a characteristic

$$\frac{(a'(\sigma))^{1/4} d\sigma}{\int_0^\sigma (a'(\eta))^{1/4} d\eta} = \frac{1}{4}\frac{\varrho'(x) dx}{\varrho(x)} ,$$

i.e.,

$$d\ln\left|\int_0^\sigma (a'(\eta))^{1/4} d\eta\right| = d\ln \varrho^{1/4}(x)$$

whence

$$\int_0^\sigma (a'(\eta))^{1/4} d\eta = C \varrho^{1/4}(x) , \qquad (2.1.16)$$

where C is constant along each characteristic.

Denote now the coordinate t for $x = 0$ by τ. Then setting $x = 0$ in (2.1.16) and taking into account the boundary condition [see (2.1.11)], we obtain

$$C = \varrho^{-1/4}(0) \int_0^{\sigma_0(\tau/\gamma)} (a'(\eta))^{1/4} d\eta .$$

Hence, (2.1.16) can be rewritten in the form

$$\int_0^\sigma (a'(\eta))^{1/4} d\eta = \left(\frac{\varrho(x)}{\varrho(0)}\right)^{1/4} \int_0^{\sigma_0(\tau/\gamma)} (a'(\eta))^{1/4} d\eta . \qquad (2.1.17)$$

On the other hand, it follows from (2.1.14) that

$$t = \tau + \int_0^x \sqrt{\varrho(z)a'(\sigma)}dz . \tag{2.1.18}$$

Formulas (2.1.17,18) give the desired solution of the problem under consideration.

Thus we have managed to analytically integrate equations of characteritics (2.1.14,15).

Problem. Estimate the coordinates of wave breaking in the problem (2.1.11,13).

2.2 When is the Factorization Exact?

2.2.1 Nonlinear Case

It is interesting to find all the cases where the factorization discrepancy $N(x,\sigma)$ is identically equal to zero. From (2.1.9) for the discrepancy

$$N(x,\sigma) = \frac{1}{4\varrho^3(x)} \int_0^\sigma \left(\frac{a'(\eta)}{a'(\sigma)}\right)^{1/4} d\eta$$
$$\times \left\{\varrho''\varrho - 2(\varrho')^2 + \frac{1}{4}(\varrho')^2 \frac{d}{d\sigma} \int_0^\sigma \left(\frac{a'(\eta)}{a'(\sigma)}\right)^{1/4} d\eta\right\} \tag{2.2.1}$$

it is clear that for $N(x,\sigma) \equiv 0$ the expression in braces on the right-hand side of (2.2.1) must be identically equal to zero. In particular, the mentioned expression in braces must not depend on σ; therefore

$$\frac{d}{d\sigma} \int_0^\sigma \left(\frac{a'(\eta)}{a'(\sigma)}\right)^{1/4} d\eta = A \tag{2.2.2}$$

whence

$$\frac{\int_0^\sigma (a'(\eta))^{1/4} d\eta}{(a'(\sigma))^{1/4}} = A\sigma + B$$

where A and B are some constants.

We consider the function $a'(\sigma)$ to be non-negative and non-vanishing identically on any interval of σ–axis. Therefore the left-hand side of the previous equality is negative for $\sigma < 0$ and positive for $\sigma > 0$. Hence, it follows that $B = 0$, $A > 0$. Thus the previous equality can be rewritten in the form

$$\frac{(a'(\sigma))^{1/4}}{\int_0^\sigma (a'(\eta))^{1/4} d\eta} = \frac{1}{A\sigma} ,$$

i.e.,

$$d\ln\left|\int_0^\sigma (a'(\eta))^{1/4} d\eta\right| = d\ln|\sigma|^{1/A}$$

whence

$$a(\sigma) = C|\sigma|^{4/A-3}\operatorname{sign}\sigma, \quad C = \text{const}. \tag{2.2.3}$$

It is clear that the power of $|\sigma|$ in (2.2.3) must be positive, whence $A < 4/3$. Thus we finally obtain that the value of A must belong to the interval

$$0 < A < 4/3. \tag{2.2.4}$$

Furthermore, from (2.2.1,2) it follows that $N(x,\sigma) \equiv 0$ if the function $\varrho(x)$ satisfies the equation

$$\varrho''\varrho - 2(\varrho')^2 + (\varrho')^2 A/4 = 0$$

or, which is the same,

$$\frac{\varrho''}{\varrho'} = \left(2 - \frac{A}{4}\right)\frac{\varrho'}{\varrho}.$$

From the last equality one can easily obtain

$$\varrho(x) = (C_1 x + C_2)^{4/(A-4)}, \tag{2.2.5}$$

$$C_1 = \text{const}, \quad C_2 = \text{const}.$$

Obviously we should assume that $C_1 x + C_2 > 0$ for x under consideration.

Thus, if the function $\varrho(x)$ is defined by (2.2.5) and the function $a(\sigma)$ by (2.2.3), then the factorization (2.1.10) of (2.1.3) turns out to be exact. It is also easy to see that (2.2.3,5) embrace all the cases of exact factorization of the nonlinear wave equation (2.1.3). (In this connection see also [2.2,3].)

2.2.2 Linear Case

Let us now consider the linear case

$$a(\sigma) = k\sigma, \quad k = \text{const} > 0$$

taking the density ϱ in accordance with (2.2.5) in the form

$$\varrho(x) = (C_1 x + C_2)^{-4/3}.$$

Then (2.1.3) takes the following form:

$$k\frac{\partial^2 \sigma}{\partial t^2} - \frac{\partial}{\partial x}(C_1 x + C_2)^{4/3}\frac{\partial \sigma}{\partial x} = 0. \tag{2.2.6}$$

From Theorem 2.1.1 and the results of Sect. 2.2.1, it follows that this equation can be factorized exactly in the form

2.2 When is the Factorization Exact?

$$0 = \left\{\sqrt{k}\frac{\partial}{\partial t} \mp \left(\frac{\partial}{\partial x}(C_1 x + C_2)^{2/3} - \frac{C_1}{3}(C_1 x + C_2)^{-1/3}\right)\right\}$$

$$\times \left\{\sqrt{k}\frac{\partial}{\partial t} \pm \left((C_1 x + C_2)^{2/3}\frac{\partial}{\partial x} + \frac{C_1}{3}(C_1 x + C_2)^{-1/3}\right)\right\}\sigma. \quad (2.2.7)$$

Using this factorization, one can easily construct the general solution of (2.2.6). In fact, consider at first the equation

$$\sqrt{k}\frac{\partial \sigma}{\partial t} + (C_1 x + C_2)^{2/3}\frac{\partial \sigma}{\partial x} + \frac{C_1}{3}(C_1 x + C_2)^{-1/3}\sigma = 0 \quad (2.2.8)$$

corresponding to the inner multiplier in (2.2.7) where upper signs are taken. Evidently, all the solutions of this equation will also be the exact solutions of (2.2.6). Let us now write the equations of characteristics for (2.2.8):

$$\frac{dt}{dx} = \sqrt{k}(C_1 x + C_2)^{-2/3}, \quad \frac{d\sigma}{dx} = -\frac{C_1}{3}(C_1 x + C_2)^{-1}\sigma.$$

Integrating, we easily obtain the general solution of these equations

$$t = \tau + \sqrt{k}\int_{x_0}^{x}(C_1 x + C_2)^{-2/3}dx,$$

$$\sigma = f(\tau)(C_1 x + C_2)^{-1/3}$$

where f is an arbitrary function. Hence the general solution of (2.2.8) has the form

$$\sigma = \frac{f\left(t - 3\sqrt{k}(C_1 x + C_2)^{1/3}/C_1\right)}{(C_1 x + C_2)^{1/3}}.$$

In a similar way we can solve the equation

$$\sqrt{k}\frac{\partial \sigma}{\partial t} - (C_1 x + C_2)^{2/3}\frac{\partial \sigma}{\partial x} - \frac{C_1}{3}(C_1 x + C_2)^{-1/3}\sigma = 0$$

corresponding to the case of lower signs in the inner multiplier in (2.2.7). Namely, its general solution has the form

$$\sigma = \frac{g\left(t + 3\sqrt{k}(C_1 x + C_2)^{1/3}/C_1\right)}{(C_1 x + C_2)^{1/3}}$$

where g is an arbitrary function.

Now, it is clear that the general solution of the linear equation (2.2.6) can be expressed by the sum:

$$\sigma = \frac{f(t - M) + (t + M)}{(C_1 x + C_2)^{1/3}},$$

$$M = 3\sqrt{k}(C_1 x + C_2)^{1/3}/C_1. \quad (2.2.9)$$

The obtained result is well-known in gas dynamics. In other variables it is given in [2.3].

2.3 Asymptotic Factorization of the General Nonlinear Wave Equation with Variable Coefficients

2.3.1 Preliminary Notes

In Sect. 2.3 we give a generalization of Theorem 2.1.1 in the case where both the nonlinear function a and the density ϱ depend on t, x:

$$a = a(t, x, \sigma), \quad \varrho = \varrho(t, x).$$

The restrictions which we impose on these functions are the following:

1) $a(t, x, \sigma)$ and $\varrho(t, x)$ are twice continuously differentiable functions of their arguments;
2) $a(t, x, 0) = 0$;
3) $a'_\sigma(t, x, \sigma) > 0$;
4) $\varrho(t, x) \geq \mathrm{const} > 0$.

The nonlinear wave equation for stress will evidently now take on the following form in the Lagrangian coordinates:

$$\frac{\partial^2 a(t, x, \sigma)}{\partial t^2} - \frac{\partial}{\partial x} \frac{1}{\varrho(t, x)} \frac{\partial \sigma}{\partial x} = 0. \tag{2.3.1}$$

Though it may seem surprising, (2.3.1), as we shall see below, can also be factorized with accuracy to the addends which do not contain derivatives of the unknown function σ.

2.3.2 Notation

Before proceeding to the factorization of (2.3.1), we make the following remark about the notation for the derivatives used in Sect. 2.3. Let

$$f = f(t, x, \sigma).$$

Then by f'_t and f'_x we denote the partial derivatives of the function f for σ fixed.

As to the derivatives with respect to t and x of the composite function $f(t, x, \sigma(t, x))$, we shall use the notation $\partial f/\partial t$ and $\partial f/\partial x$. Thus, from the theorem about the derivative of a composite function we have

$$\frac{\partial f}{\partial t} = f'_t + f'_\sigma \frac{\partial \sigma}{\partial t}, \quad \text{and} \quad \frac{\partial f}{\partial x} = f'_x + f'_\sigma \frac{\partial \sigma}{\partial x}.$$

2.3.3 Representation of the General Nonlinear Wave Equation with Variable Coefficients

Let us seek the representation of the nonlinear operator from (2.3.1) in the form

$$\frac{\partial^2 a(t,x,\sigma)}{\partial t^2} - \frac{\partial}{\partial x} \frac{1}{\varrho(t,x)} \frac{\partial \sigma}{\partial x}$$
$$\equiv \left\{ \frac{\partial}{\partial t} \sqrt{a'_\sigma(t,x,\sigma)} - \frac{\partial}{\partial x} \frac{1}{\sqrt{\varrho(t,x)}} + \varphi_1(t,x,\sigma) \right\}$$
$$\times \left\{ \sqrt{a'_\sigma(t,x,\sigma)} \frac{\partial \sigma}{\partial t} + \frac{1}{\sqrt{\varrho(t,x)}} \frac{\partial \sigma}{\partial x} + \varphi_2(t,x,\sigma) \right\}$$
$$- N(t,x,\sigma) .$$
(2.3.2)

For simplicity we again restrict ourselves to the case where the inner multiplier in the decomposition describes waves travelling to the right. Multiplying out the braces on the right-hand side of (2.3.2), we get the following expression:

$$\frac{\partial}{\partial t} a'_\sigma \frac{\partial \sigma}{\partial t} - \frac{\partial}{\partial x} \sqrt{\frac{a'_\sigma}{\varrho}} \frac{\partial \sigma}{\partial t} + \varphi_1 \sqrt{a'_\sigma} \frac{\partial \sigma}{\partial t} + \frac{\partial}{\partial t} \sqrt{\frac{a'_\sigma}{\varrho}} \frac{\partial \sigma}{\partial x}$$
$$- \frac{\partial}{\partial x} \frac{1}{\varrho} \frac{\partial \sigma}{\partial x} + \varphi_1 \frac{1}{\sqrt{\varrho}} \frac{\partial \sigma}{\partial x} + \frac{\partial}{\partial t} \sqrt{a'_\sigma} \varphi_2 - \frac{\partial}{\partial x} \frac{1}{\sqrt{\varrho}} \varphi_2$$
$$+ \varphi_1 \varphi_2 - N(t,x,\sigma) .$$
(2.3.3)

Before reducing similar terms in (2.3.3) note that the following identities hold:

$$\frac{\partial}{\partial t} a'_\sigma \frac{\partial \sigma}{\partial t} \equiv \frac{\partial^2 a}{\partial t^2} - a''_{tt} - a''_{t\sigma} \frac{\partial \sigma}{\partial t}$$
(2.3.4)

and

$$-\frac{\partial}{\partial x} \sqrt{\frac{a'_\sigma}{\varrho}} \frac{\partial \sigma}{\partial t} + \frac{\partial}{\partial t} \sqrt{\frac{a'_\sigma}{\varrho}} \frac{\partial \sigma}{\partial x}$$
$$\equiv -\left(\sqrt{\frac{a'_\sigma}{\varrho}}\right)'_x \frac{\partial \sigma}{\partial t} + \left(\sqrt{\frac{a'_\sigma}{\varrho}}\right)'_t \frac{\partial \sigma}{\partial x} .$$
(2.3.5)

The identities (2.3.4,5) can be verified by direct differentiation. Now, taking into account (2.3.4,5), let us rewrite (2.3.3) as

2. Nonlinear Short Waves of Finite Amplitude

$$\frac{\partial^2 a}{\partial t^2} - \frac{\partial}{\partial x}\frac{1}{\varrho}\frac{\partial \sigma}{\partial x}$$

$$+ \left\{ \varphi_1 \sqrt{a'_\sigma} + \left(\sqrt{a'_\sigma}\right)'_\sigma \varphi_2 + \sqrt{a'_\sigma}\, (\varphi_2)'_\sigma - \left(\sqrt{\frac{a'_\sigma}{\varrho}}\right)'_x - a''_{t\sigma} \right\} \frac{\partial \sigma}{\partial t}$$

$$+ \left\{ \varphi_1 \frac{1}{\sqrt{\varrho}} - \frac{1}{\sqrt{\varrho}}(\varphi_2)'_\sigma + \left(\sqrt{\frac{a'_\sigma}{\varrho}}\right)'_t \right\} \frac{\partial \sigma}{\partial x}$$

$$+ \left\{ -a''_{tt} + \left(\sqrt{a'_\sigma}\right)'_t \varphi_2 + \sqrt{a'_\sigma}(\varphi_2)'_t - \left(\frac{1}{\sqrt{\varrho}}\right)'_x \varphi_2 \right.$$

$$\left. - \frac{1}{\sqrt{\varrho}}(\varphi_2)'_x + \varphi_1 \varphi_2 \right\} - N(t,x,\sigma). \tag{2.3.6}$$

To obtain the desired representation of the nonlinear wave operator under consideration, let us equate the coefficients to $\partial\sigma/\partial t$ and $\partial\sigma/\partial x$ with zero:

$$\varphi_1 \sqrt{a'_\sigma} + \left(\sqrt{a'_\sigma}\right)'_\sigma \varphi_2 + \sqrt{a'_\sigma}(\varphi_2)'_\sigma - \left(\sqrt{\frac{a'_\sigma}{\varrho}}\right)'_x - a''_{t\sigma} = 0; \tag{2.3.7}$$

$$\frac{1}{\sqrt{\varrho}}\varphi_1 - \frac{1}{\sqrt{\varrho}}(\varphi_2)'_\sigma + \left(\sqrt{\frac{a'_\sigma}{\varrho}}\right)'_t = 0. \tag{2.3.8}$$

Note now that in (2.3.7,8) the variables t, x are, in fact, parameters; therefore the system itself can be considered as a system of *ordinary* differential equations where σ is playing the role of the independent variable.

By eliminating φ_1 from (2.3.7,8), we obtain the following equation for φ_2:

$$2\sqrt{a'_\sigma}\frac{d\varphi_2}{d\sigma} + \left(\sqrt{a'_\sigma}\right)'_\sigma \varphi_2 = \left(\sqrt{\frac{a'_\sigma}{\varrho}}\right)'_x + \sqrt{\varrho a'_\sigma}\left(\sqrt{\frac{a'_\sigma}{\varrho}}\right)'_t + a''_{t\sigma},$$

i.e.,

$$2(a'_\sigma)^{1/4}\frac{d}{d\sigma}\left\{(a'_\sigma)^{1/4}\varphi_2\right\} = \left(\sqrt{\frac{a'_\sigma}{\varrho}}\right)'_x + \sqrt{\varrho a'_\sigma}\left(\sqrt{\frac{a'_\sigma}{\varrho}}\right)'_t + a''_{t\sigma}.$$

Hence,

$$\varphi_2 = \frac{1}{2}(a'_\sigma)^{-1/4}\int_0^\sigma (a'_\sigma)^{-1/4}\, Q\, d\sigma,$$

$$Q = \left(\sqrt{\frac{a'_\sigma}{\varrho}}\right)'_x + \sqrt{\varrho a'_\sigma}\left(\sqrt{\frac{a'_\sigma}{\varrho}}\right)'_t + a''_{t\sigma}. \tag{2.3.9}$$

2.3 Asymptotic Factorization of General Nonlinear Wave Equation

We set here the lower limit of integration equal to zero to satisfy the equality $\varphi_2 = 0$ for $\sigma = 0$.

Now from (2.3.8,9) we find φ_1

$$\varphi_1 = \frac{1}{2}\left\{(a'_\sigma)^{-1/4}\int_0^\sigma (a'_\sigma)^{-1/4} Q\, d\sigma\right\}'_\sigma - \sqrt{\varrho}\left(\sqrt{\frac{a'_\sigma}{\varrho}}\right)'_t. \qquad (2.3.10)$$

After the functions φ_1 and φ_2 have been found, we simply put the function $N(t, x, \sigma)$ equal to the contents of the last braces in (2.3.6), which yields the representation (2.3.2). In the same way it is easy to obtain a representation, analogous to (2.3.2), in which the inner multiplier in the product of braces describes waves travelling to the left.

Thus we arrive at the following result.

Theorem 2.3.3. [2.1] The equation (2.3.1) can be represented in the form

$$\left\{\frac{\partial}{\partial t}\sqrt{a'_\sigma(t,x,\sigma)} \mp \frac{\partial}{\partial x}\frac{1}{\sqrt{\varrho(t,x)}} + \varphi_1(t,x,\sigma)\right\}$$
$$\times \left\{\sqrt{a'_\sigma(t,x,\sigma)}\frac{\partial \sigma}{\partial t} \pm \frac{1}{\sqrt{\varrho(t,x)}}\frac{\partial \sigma}{\partial x} + \varphi_2(t,x,\sigma)\right\}$$
$$= N(t, x, \sigma) \qquad (2.3.11)$$

where

$$\varphi_1 = \frac{1}{2}\left\{(a'_\sigma)^{-1/4}\int_0^\sigma (a'_\sigma)^{-1/4}\left[\pm\left(\sqrt{\frac{a'_\sigma}{\varrho}}\right)'_x\right.\right.$$
$$\left.\left.+ \sqrt{\varrho a'_\sigma}\left(\sqrt{\frac{a'_\sigma}{\varrho}}\right)'_t + a''_{t\sigma}\right]d\sigma\right\}'_\sigma - \sqrt{\varrho}\left(\sqrt{\frac{a'_\sigma}{\varrho}}\right)'_t; \qquad (2.3.12)$$

$$\varphi_2 = \frac{1}{2}(a'_\sigma)^{-1/4}\int_0^\sigma (a'_\sigma)^{-1/4}\left[\pm\left(\sqrt{\frac{a'_\sigma}{\varrho}}\right)'_x\right.$$
$$\left.+ \sqrt{\varrho a'_\sigma}\left(\sqrt{\frac{a'_\sigma}{\varrho}}\right)'_t + a''_{t\sigma}\right]d\sigma \qquad (2.3.13)$$

and the difference $N(t, x, \sigma)$ is given by

$$N = -a''_{tt} + \left(\sqrt{a'_\sigma}\right)'_t \varphi_2 + \sqrt{a'_\sigma}\,(\varphi_2)'_t$$
$$\mp \left(\frac{1}{\sqrt{\varrho}}\right)'_x \varphi_2 \mp \frac{1}{\sqrt{\varrho}}(\varphi_2)'_x + \varphi_1\varphi_2. \qquad (2.3.14)$$

In (2.3.11–14) either upper or lower signs are simultaneously chosen.

Remark 1. In (2.3.13) for the function φ_2 we have chosen the lower limit of integration equal to zero. Hence, it follows that the discrepancy $N(t, x, \sigma)$ vanishes for $\sigma = 0$ (since $a(t, x, 0) = 0$). Thus, the chosen form of representation of (2.3.1) is adapted for investigation of pulse propagation in the undisturbed medium.

Remark 2. One can easily see that for $a = a(\sigma)$, $\varrho = \varrho(x)$ the previous theorem turns into Theorem 2.1.1.

Problem. Under which conditions is (2.3.11) exact? (In this connection see [2.2,4].)

2.3.4 Formulation of the Boundary Value Problem Conditions of Asymptotic Factorization

For (2.3.1) let us set the problem about the short pulse propagation as follows:

$$\sigma = \frac{\partial \sigma}{\partial t} = 0 \quad \text{for} \quad t = 0, \quad x > 0,$$

$$\sigma(t, 0) = \sigma_0\left(\frac{t}{\gamma}\right), \tag{2.3.15}$$

where $\sigma_0(\tau)$ is a smooth function equal to zero for $\tau < 0$ and for $\tau > T > 0$; γ is a small parameter ($0 < \gamma \ll 1$).

We shall suppose that in (2.3.1) the characteristic scale of spatial inhomogeneity is equal to L, and the characteristic scale of time inhomogeneity is equal to M. Then, in the manner of Sect. 2.1.1, it is easy to verify that for the short pulse (2.3.15) the ratio of the discrepancy $N(t, x, \sigma)$ to the terms of (2.3.1) has the order of value $O(\gamma^2)$.

Therefore, when solving (2.3.15), one can neglect the discrepancy $N(t, x, \sigma)$ and instead of the nonlinear wave equation (2.3.1) consider the single wave equation

$$\sqrt{a'(t, x, \sigma)}\,\frac{\partial \sigma}{\partial t} + \frac{1}{\sqrt{\varrho(t, x)}}\frac{\partial \sigma}{\partial x} + \varphi_2(t, x, \sigma) = 0 \tag{2.3.16}$$

which corresponds to the choice of the upper signs in (2.3.11,13). Evidently, one can solve this equation by the method of characteristics:

$$\frac{dt}{dx} = \sqrt{\varrho(t, x) a'_\sigma(t, x, \sigma)},$$

$$\frac{d\sigma}{dx} = -\sqrt{\varrho(t, x)}\, \varphi_2(t, x, \sigma). \tag{2.3.17}$$

The system of ordinary equations (2.3.17), generally speaking, is not analytically integrable (unlike the similar system in Sect. 2.1.3). However, it is much easier to deal with such a system than with (2.3.1).

Problem. Consider a semi-infinite compound rod with the following constitutive equation and density:

$$\left.\begin{array}{l}\varepsilon = A\sigma, \quad A = \text{const} \\ \varrho = \varrho_0 = \text{const}\end{array}\right\} \quad \text{for} \quad 0 < x < l,$$

$$\left.\begin{array}{l}\varepsilon = a(t, x, \sigma) \\ \varrho = \varrho(t, x)\end{array}\right\} \quad \text{for} \quad x > l.$$

Let the rod be subjected to the load (2.3.15). Find the asymptotic expression for the wave reflected from the bound $x = l$ into the linear elastic part of the rod.

2.3.5 Linear Case

Now let

$$a = A(t, x)\sigma.$$

Then Theorem 2.3.3 yields the following result:

Theorem 2.3.5. The equation

$$\frac{\partial^2 A(t,x)\sigma}{\partial t^2} - \frac{\partial}{\partial x}\frac{1}{\varrho(t,x)}\frac{\partial \sigma}{\partial x} = 0 \qquad (2.3.18)$$

can be represented in the form

$$\left\{\frac{\partial}{\partial t}\sqrt{A} \mp \frac{\partial}{\partial x}\frac{1}{\sqrt{\varrho}} + \frac{A'_t \pm \left(\sqrt{A/\varrho}\right)'_x - \sqrt{\varrho A}\left(\sqrt{A/\varrho}\right)'_t}{2\sqrt{A}}\right\}$$

$$\times \left\{\sqrt{A}\frac{\partial \sigma}{\partial t} \pm \frac{1}{\sqrt{\varrho}}\frac{\partial \sigma}{\partial x} + \sigma\frac{A'_t \pm \left(\sqrt{A/\varrho}\right)'_x + \sqrt{\varrho A}\left(\sqrt{A/\varrho}\right)'_t}{2\sqrt{A}}\right\}$$

$$= N(t, x, \sigma) \qquad (2.3.19)$$

where the function $N(t, x, \sigma)$ is given by (2.3.14) in which $a(t, x, \sigma)$ should be replaced by $A(t, x)\sigma$. (In particular, $N(t, x, 0) = 0$.)

Remark. Theorem 2.3.5 yields two single-wave equations

$$\sqrt{A}\frac{\partial \sigma}{\partial t} \pm \frac{1}{\sqrt{\varrho}}\frac{\partial \sigma}{\partial x} + \sigma\frac{A'_t \pm \left(\sqrt{A/\varrho}\right)'_x + \sqrt{\varrho A}\left(\sqrt{A/\varrho}\right)'_t}{2\sqrt{A}} = 0,$$

corresponding to the inner multiplier on the left-hand side of (2.3.19) for the two different choices of signs. One can evidently solve these equations by the

2.4 Evolution of Maximal Amplitude of the Stress Wave

2.4.1 Formulation of the Problem

Consider now the problem (2.3.15,16) in a simpler case where the functions a and ϱ do not depend on t. Then the single-wave equation (2.3.16) which describes stress waves propagating to the right in the undisturbed medium, takes the form

$$\sqrt{a'_\sigma(x,\sigma)}\,\frac{\partial \sigma}{\partial t} + \frac{1}{\sqrt{\varrho(x)}}\,\frac{\partial \sigma}{\partial x} + \varphi_2(x,\sigma) = 0 \qquad (2.4.1)$$

where

$$\varphi_2(x,\sigma) = \frac{1}{2}\left(a'_\sigma\right)^{-1/4} \int_0^\sigma \left(a'_\sigma\right)^{-1/4} \left(\sqrt{\frac{a'_\sigma}{\varrho}}\right)'_x \, d\sigma \, . \qquad (2.4.2)$$

Let us write once more the boundary condition (2.3.15), i.e.,

$$\sigma(t,0) = \sigma_0\left(\frac{t}{\gamma}\right), \quad 0 < \gamma \ll 1 \, . \qquad (2.4.3)$$

Now, we shall impose some stronger restrictions on the boundary function. Namely, we shall suppose $\sigma_0(\tau)$ to be a smooth nonnegative function, equal to zero for $\tau < 0$ and for $\tau > T > 0$, achieving its maximum for $\tau = \tau_m$ and monotone on the intervals $(0, \tau_m)$ and (τ_m, T).

Furthermore, denote by $t(x)$ the value of t for which the stress achieves its maximum (for x fixed), i.e.,

$$\max_t \sigma(t,x) = \sigma(t(x), x) \, . \qquad (2.4.4)$$

Thus, the line $t = t(x)$ is a curve along which the stress maximum propagates. It is intuitively clear that under the assumptions made above the function $t = t(x)$ will be single-valued.

2.4.2 Equation for Maximal Amplitudes

Let us now try to describe the evolution of the stress maximal amplitude (2.4.4) dependent on x. Denote

$$w(x) = \sigma(t(x), x) \, . \qquad (2.4.5)$$

It is clear that the function w introduced above satisfies the condition

$$w(0) = \sigma_0(\tau_m) . \tag{2.4.6}$$

Furthermore, note that by application of the definition of the curve $t = t(x)$ as the line of maximums, the following relation must hold:

$$\left.\frac{\partial \sigma}{\partial t}\right|_{t=t(x)} = 0 .$$

Therefore, it is evident that

$$\left.\frac{\partial \sigma}{\partial x}\right|_{t=t(x)} = \frac{dw}{dx} .$$

Substituting $t = t(x)$ into (2.4.1) and taking into account the two previous relations, we obtain an ordinary differential equation for $w(x)$

$$\frac{dw}{dx} = -\sqrt{\varrho(x)}\varphi_2(x, w) . \tag{2.4.7}$$

The equation (2.4.7), which was derived by Lokshin and Enikeeva, is the desired ordinary differential equation describing the maximal stress evolution. The initial condition for this equation is given by (2.4.6).

2.4.3 The Curve of Maximums as a Characteristic

Note, however, that from (2.4.7) it is impossible to determine the moment at which the stress maximum is achieved (for $x = $ const). That is, the curve of maximums has not yet been obtained.

Nevertheless, one can find the curve $t = t(x)$ by the following reasoning. Let us write the characteristics for (2.4.1) as

$$\frac{dt}{dx} = \sqrt{\varrho(x) a'_\sigma(x, \sigma)} , \tag{2.4.8}$$

$$\frac{d\sigma}{dx} = -\sqrt{\varrho(x)}\varphi_2(x, \sigma) \tag{2.4.9}$$

and consider the characteristic starting from the point $t = \gamma\tau_m$ of the t-axis. It is clear that for $x = 0$ we have $\sigma = \sigma_0(\tau_m)$ on this characteristic. Note now that (2.4.9) coincides with (2.4.7). Since the initial conditions for these two equations also coincide, then along the characteristic considered

$$\sigma(t, x) = w(x) .$$

Hence, it evidently follows that the characteristic under consideration must coincide with the curve of maximums $t = t(x)$. Thus, after the function $w(x)$ has been obtained from (2.4.6,7), the curve $t = t(x)$ will be determined as the solution of

$$\frac{dt}{dx} = \sqrt{\varrho(x) a'_\sigma(x, w(x))} ; \quad t(0) = \gamma\tau_m . \tag{2.4.10}$$

Remark. In a similar way one can investigate the case where $\sigma_0(\tau)$ is a smooth nonpositive function equal to zero for $\tau < 0$ and for $\tau > T > 0$, achieving its minimum for $\tau = \tilde{\tau}_m$ and monotone on the intervals $(0, \tilde{\tau}_m)$ and $(\tilde{\tau}_m, T)$.

Finally, of interest is the case where the boundary function $\sigma_0(\tau)$ vanishing outside $[0, T]$ has several maximums and minimums. It should be natural to expect that all these stress maximums and minimums will also travel along the corresponding characteristics. The validity of this hypothesis for the more special equation (2.1.13) is evident from (2.1.17,18).

2.5 Propagation of a Stress Wave in a Homogeneous Nonlinear Elastic Rod Located in the Gravity Field

2.5.1 Formulation of the Problem

Now we shall explore one more application of Theorem 2.3.3. Consider a rod which in the nondeformed state is homogeneous, has the density ϱ and the length L. We suppose that in this rod deformation ε and stress σ are related by the constitutive equation

$$\varepsilon = a(\sigma), \qquad (2.5.1)$$

where $a'(\sigma) > 0$, $a(0) = 0$.

Furthermore, let the rod under consideration be located along the vertical axis Ox (which is directed downwards) in the homogeneous field of gravity and occupy in the Lagrangian coordinates (connected with the nondeformed state) the segment $0 \leq x \leq L$. The equations of motion of the rod in the Lagrangian coordinates are evidently the following:

$$\frac{\partial v}{\partial t} = \frac{1}{\varrho} \frac{\partial \sigma}{\partial x} + g, \quad \text{and} \quad \frac{\partial v}{\partial x} = \frac{\partial \varepsilon}{\partial t} \qquad (2.5.2)$$

where g is the gravitational acceleration. Eliminating v from (2.5.2) and using the constitutive equation (2.5.1), we arrive at the well-known nonlinear wave equation with constant coefficients

$$\frac{\partial^2 a(\sigma)}{\partial t^2} - \frac{1}{\varrho} \frac{\partial^2 \sigma}{\partial x^2} = 0. \qquad (2.5.3)$$

Now, we discover some new properties of this equation.

We suppose that for $t = 0$ the following initial conditions hold:

$$v(0, x) = \frac{\partial v(0, x)}{\partial t} = 0.$$

Then from (2.5.1,2) we immediately obtain the initial conditions for (2.5.3)

$$\sigma(0,x) = -\varrho g x + \text{const}; \quad \frac{\partial \sigma(0,x)}{\partial t} = 0.$$

To be definite, we consider the lower end of the rod ($x = L$) to be free, i.e.,

$$\sigma(t, L) = 0. \tag{2.5.4}$$

Applying this, the initial conditions for (2.5.3) finally take the form

$$\sigma(0,x) = \varrho g(L-x); \quad \frac{\partial \sigma(0,x)}{\partial t} = 0. \tag{2.5.5}$$

Let us now set the boundary condition for (2.5.3) on the upper end of the rod ($x = 0$) as follows:

$$\sigma(t,0) = \varrho g L + \sigma_0(t/\gamma) \tag{2.5.6}$$

where $\sigma_0(\tau)$ is a smooth function equal to zero for $\tau < 0$ and for $\tau > T > 0$; $0 < \gamma \ll 1$.

Remark. We shall consider the problem only for the moments of time preceding the pulse reflection from the free end of the rod. Thus, the setting of the problem is, in fact, determined by (2.5.3), the initial conditions (2.5.5) and the boundary condition (2.5.6).

2.5.2 Uselessness of Exact Factorization

As we know from Theorem 1.3.2, the equation (2.5.3) can be represented by

$$\left\{ \frac{\partial}{\partial t} \sqrt{a'(\sigma)} \mp \frac{1}{\sqrt{\varrho}} \frac{\partial}{\partial x} \right\} \left\{ \sqrt{a'(\sigma)} \frac{\partial \sigma}{\partial t} \pm \frac{1}{\sqrt{\varrho}} \frac{\partial \sigma}{\partial x} \right\} = 0 \tag{2.5.7}$$

where either upper or lower signs are simultaneously chosen. Consider the equation

$$\sqrt{a'(\sigma)} \frac{\partial \sigma}{\partial t} + \frac{1}{\sqrt{\varrho}} \frac{\partial \sigma}{\partial x} = 0 \tag{2.5.8}$$

corresponding to the inner multiplier in (2.5.7) if we choose the upper signs. Evidently this equation describes waves travelling to the right. However, (2.5.8) is unsuitable for our problem, because none of its solutions can simultaneously satisfy both initial and boundary conditions (2.5.5,6). (One can easily demonstrate this fact by the method of characteristics.)

2.5.3 Asymptotic Factorization

Nevertheless, using Theorem 2.3.3, we shall be able to derive a single-wave equation describing stress wave propagation in (2.5.3,5,6).

It is quite natural to introduce the following change of the unknown function:

$$\sigma = \varrho g(L - x) + w(t, x) \, . \tag{2.5.9}$$

Then (2.5.3) will take the form

$$\frac{\partial^2}{\partial t^2} a(w + \varrho g(L - x)) - \frac{1}{\varrho} \frac{\partial^2 w}{\partial x^2} = 0 \, . \tag{2.5.10}$$

But this equation evidently belongs to the class of equations to which Theorem 2.3.3 is applicable. In the special case of (2.5.10) Theorem 2.3.3 can be formulated in the following way:

Theorem 2.5.3. Equation (2.5.10) can be represented in the form

$$N(x, w) = \left\{ \frac{\partial}{\partial t} \sqrt{a'(w + \varrho g(L - x))} \mp \frac{1}{\sqrt{\varrho}} \left(\frac{\partial}{\partial x} - f'_w(x, w) \right) \right\}$$
$$\times \left\{ \sqrt{a'(w + \varrho g(L - x))} \frac{\partial w}{\partial t} \pm \frac{1}{\sqrt{\varrho}} \left(\frac{\partial w}{\partial x} + f(x, w) \right) \right\} \tag{2.5.11}$$

where

$$f(x, w) = \varrho g \left\{ \left(\frac{a'(\varrho g(L - x))}{a'(w + \varrho g(L - x))} \right)^{1/4} - 1 \right\} \tag{2.5.12}$$

and

$$N(x, w) = \frac{1}{\varrho} \left\{ \mp f'_x(x, w) + f(x, w) f'_w(x, w) \right\} \, . \tag{2.5.13}$$

(In (2.5.11,13) either upper or lower signs are simultaneously chosen; the derivative f'_x in (2.5.13) is taken for w fixed.)

Remark. It is clear that the difference (2.5.13) vanishes for $w = 0$.

One can easily see that for the short pulse determined by the boundary condition (2.5.6) the ratio of the difference $N(x, w)$ to the terms of the nonlinear wave equation (2.5.10) will be of order $O(\gamma^2)$.

Neglecting this discrepancy and returning from w to the variable σ, we obtain from (2.5.11) that the nonlinear wave equation (2.5.3) for the short pulse under consideration can be asymptotically represented as a product:

2.5 Propagation of a Stress Wave 53

$$\left\{ \frac{\partial}{\partial t} \sqrt{a'(\sigma)} \mp \frac{1}{\sqrt{\varrho}} \left(\frac{\partial}{\partial x} - \varrho g \frac{d}{d\sigma} \left(\frac{a'(\varrho g(L-x))}{a'(\sigma)} \right)^{1/4} \right) \right\}$$
$$\times \left\{ \sqrt{a'(\sigma)} \frac{\partial \sigma}{\partial t} \pm \frac{1}{\sqrt{\varrho}} \left(\frac{\partial \sigma}{\partial x} + \varrho g \left(\frac{a'(\varrho g(L-x))}{a'(\sigma)} \right)^{1/4} \right) \right\} = 0 \quad (2.5.14)$$

[the corresponding discrepancy vanishes for $\sigma = \varrho g(L - x)$].

2.5.4 Single-Wave Solution of the Problem

Now to asymptotically solve (2.5.3,5,6), we can use the first order equation corresponding to the inner multiplier in the decomposition (2.5.14), where the upper signs are chosen, i.e.,

$$\sqrt{\varrho a'(\sigma)} \frac{\partial \sigma}{\partial t} + \frac{\partial \sigma}{\partial x} + \varrho g \left(\frac{a'(\varrho g(L-x))}{a'(\sigma)} \right)^{1/4} = 0. \quad (2.5.15)$$

Let us write the equation of characteristics for (2.5.15) as

$$\frac{dt}{dx} = \sqrt{\varrho a'(\sigma)},$$

$$\frac{d\sigma}{dx} = -\varrho g \left(\frac{a'(\varrho g(L-x))}{a'(\sigma)} \right)^{1/4},$$

whence, by integrating under the boundary condition (2.5.6), we obtain

$$\int_{\varrho g L + \sigma_0(\tau/\gamma)}^{\sigma} \left(a'(\sigma) \right)^{1/4} d\sigma = \int_{\varrho g L}^{\varrho g (L-x)} \left(a'(\xi) \right)^{1/4} d\xi, \quad (2.4.16)$$

$$t = \tau + \int_0^x \sqrt{\varrho a'(\sigma)} dx. \quad (2.5.17)$$

In (2.5.17) σ is determined by (2.5.16) as a function of τ and x.

One can easily see that the constructed above solution (2.5.16,17) satisfies not only the boundary condition (2.5.6) but also the initial conditions (2.5.5).

Thus, the stated problem (2.5.3,5,6) is asymptotically solved. Note, finally, that in the special case where $a(\sigma)$ is a linear function our solution is exact.

Problem. Let

$$a(\sigma) = A\sigma, \quad A = \text{const}$$

for $|\sigma| \leq \text{const}$. Moreover, suppose that near the free end of the rod the incident wave amplitude is small enough for the incident and reflected waves to have no interaction. Find the reflected wave.

3. Nonlinear Waves in Media with Memory

Chapter 3 deals with nonlinear one-dimensional waves in hereditary elastic media, i.e., the ones in which the relation between stress and deformation depends on the entire history of the process. In the wave equation describing, for example, stress propagation, the dependence on the history of the process is given by an integral in time. Characteristics are usually useless for deriving the solutions of these equations. However, there are important exceptions which will be discussed below.

The basic tool is still the factorization of wave equations.

3.1 Hereditary Elasticity

3.1.1 Linear Equations

Let us consider a homogeneous rod made of a linear hereditary elastic material. In such a rod the deformation ε and the stress σ are related by the constitutive equation [3.1–4]

$$\varepsilon(t) = \frac{1}{E}\left(\sigma(t) + \int_{-\infty}^{t} K(t-\tau)\sigma(\tau)d\tau\right) \tag{3.1.1}$$

which can also be written

$$\sigma(t) = E\left(\varepsilon(t) - \int_{-\infty}^{t} R(t-\tau)\varepsilon(\tau)d\tau\right) . \tag{3.1.2}$$

Here E is the *instantaneous modulus of elasticity*, $K(t)$ is the *creep kernel*, $R(t)$ is the *relaxation kernel*. The relation of the creep kernel to the relaxation kernel is given by

$$\begin{aligned}K(t) = \ & R(t) + \int_0^t R(t-\tau)R(\tau)d\tau \\ & + \int_0^t R(t-\tau)\int_0^\tau R(\tau-\tau_1)R(\tau_1)d\tau_1\,d\tau + \ldots\end{aligned} \tag{3.1.3}$$

where the dots denote multiple convolutions of higher orders.

It is clear that for (3.1.1,2) to make sense it is necessary that $K(t)$ and $R(t)$, which are defined on the semi-axis $t > 0$, should be integrable on $[0,\infty)$.

In hereditary elasticity the kernels $K(t)$ and $R(t)$ are usually considered nonnegative and decreasing. This assumption is called the *principle of fading memory*. (Note, however, that hereditary kernels for vapor-liquid and bubble media turn out to be oscillatory functions [3.5,6].)

It is clear from (3.1.3) that $K(t) \sim R(t)$ as $t \to 0$. The functions $K(t)$, $R(t)$ are called *regular* if they tend to a finite limit as $t \to +0$, and *singular* if they tend to $+\infty$ as $t \to +0$.

3.1.2 Nonlinear Equations

If deformations are not very small, then it is usually necessary to introduce nonlinearity into (3.1.1,2). Two of the best known nonlinear hereditary models are the *Rabotnov model*

$$\varepsilon(t) = a\left(\sigma(t) + \int_{-\infty}^{t} K(t-\tau)\sigma(\tau)d\tau\right)$$

where one nonlinear function, a, is used, and the *Leaderman–Rosovski model*

$$\varepsilon(t) = a(\sigma(t)) + \int_{-\infty}^{t} K(t-\tau)b(\sigma(\tau))d\tau$$

where two nonlinear functions, a and b, are used.

Later on we shall use the special case of the Leaderman–Rosovski equation where

$$\varepsilon(t) = a(\sigma(t)) + \int_{-\infty}^{t} K(t-\tau)a(\sigma(\tau))d\tau \ . \tag{3.1.4}$$

This equation can also be written as

$$\varepsilon(t) - \int_{-\infty}^{t} R(t-\tau)\varepsilon(\tau)d\tau = a(\sigma(t)) \tag{3.1.4'}$$

where the kernel $R(t)$ is still related to $K(t)$ by (3.1.3).

If $\varepsilon = \sigma = 0$ for $t < 0$, then it follows from both the Rabotnov and the Leaderman–Rosovski models that

$$\varepsilon(t) \sim a(\sigma(t)) \quad \text{as} \quad t \to +0 \ . \tag{3.1.5}$$

Moreover, if loading is fast enough for $t > 0$, then (3.1.5) will hold even for large values of the moduli of ε and σ. Relation (3.1.5) is sometimes called the *curve of instantaneous deformation*.

For convenience we introduce the following notation for the convolution operators:

$$K*u \equiv \int_{-\infty}^{t} K(t-\tau)u(\tau)d\tau \ ,$$

$$R^*u \equiv \int_{-\infty}^{t} R(t-\tau)u(\tau)d\tau$$

where u is a test function. In this notation (3.1.4') becomes

$$\varepsilon = \frac{1}{1-R^*} a(\sigma), \tag{3.1.6}$$

where $1/(1-R^*)$ means $1 + R^* + R^*R^* + \ldots$. Note also that (3.1.3) yields the following relation between the operators K^* and R^*:

$$1 + K^* = \frac{1}{1-R^*}.$$

Indeed, the Rabotnov and Leaderman–Rosovski models do not embrace the whole variety of constitutive equations with the curve of instantaneous deformation $\varepsilon \sim a(\sigma)$. For example, the same curve of instantaneous deformation corresponds to the constitutive equation

$$\varepsilon(t) = \int_{-\infty}^{t} \sqrt{1+K^*}\sqrt{a'(\sigma)}\sqrt{1+K^*}\sqrt{a'(\sigma)}\sigma'_t dt \tag{3.1.7}$$

where it is assumed that $\sigma \to 0$ as $t \to -\infty$. In this constitutive equation

$$\sqrt{1+K^*} = 1 + \frac{K^*}{2} + \ldots$$

and the expression

$$\sqrt{1+K^*}\sqrt{a'(\sigma)}\sqrt{1+K^*}\sqrt{a'(\sigma)}$$

must be regarded as a product of operators (applied to σ'_t). In addition it is assumed that the operators in the product act in turn from right to left. Although this constitutive equation seems rather complicated, one can find it advantageous when solving wave problems.

Remark. All the algebraic and rational functions of the convolution operators $K^*, R^*, \ldots$ which appear below, are considered as Taylor series in powers of these operators.

3.2 Small Quadratic Nonlinearity

3.2.1 Asymptotic Factorization of the Nonlinear Wave Equation with Memory

Suppose that in a homogeneous rod of density ϱ stress and deformation are related by the constitutive equation

$$\varepsilon = \frac{1}{1-\gamma R^*} \left(A\sigma + \gamma B\sigma^2\right), \quad 0 < \gamma \ll 1 \tag{3.2.1}$$

where

$$A > 0, \quad B/A^2 = O(1), \quad \int_0^\infty R(t)dt = O(1).$$

We write the equations of motion of the rod in the Lagrangian coordinates

$$\frac{1}{\varrho}\frac{\partial \sigma}{\partial x} = \frac{\partial v}{\partial t}, \quad \frac{\partial v}{\partial x} = \frac{\partial \varepsilon}{\partial t}.$$

These equations, as we know, yield the equality

$$\frac{\partial^2 \varepsilon}{\partial t^2} - \frac{1}{\varrho}\frac{\partial^2 \sigma}{\partial x^2} = 0. \tag{3.2.2}$$

Substituting (3.2.1) into the previous equality, we obtain a nonlinear wave equation with memory

$$\frac{\partial^2 (A\sigma + \gamma B\sigma^2)}{\partial t^2} - \frac{1}{\varrho}(1 - \gamma R^*)\frac{\partial^2 \sigma}{\partial x^2} = 0. \tag{3.2.3}$$

Theorem 3.2.1 [3.7]. Equation (3.2.3) can be asymptotically factorized in the following way:

$$\left\{\frac{\partial}{\partial t}\sqrt{A + 2\gamma B\sigma} \mp \sqrt{\frac{1-\gamma R^*}{\varrho}}\frac{\partial}{\partial x}\right\}$$

$$\times \left\{\sqrt{A + 2\gamma B\sigma}\frac{\partial}{\partial t} \pm \sqrt{\frac{1-\gamma R^*}{\varrho}}\frac{\partial}{\partial x}\right\} \sigma = O(\gamma^2). \tag{3.2.4}$$

Proof. Multiplying out the expressions in braces in (3.2.4), we obtain

$$\left\{\frac{\partial}{\partial t}\sqrt{A + 2\gamma B\sigma} \mp \sqrt{\frac{1-\gamma R^*}{\varrho}}\frac{\partial}{\partial x}\right\}\left\{\sqrt{A + 2\gamma B\sigma}\frac{\partial \sigma}{\partial t} \pm \sqrt{\frac{1-\gamma R^*}{\varrho}}\frac{\partial \sigma}{\partial x}\right\}$$

$$\equiv \frac{\partial}{\partial t}\sqrt{A+2\gamma B\sigma}\sqrt{A+2\gamma B\sigma}\frac{\partial \sigma}{\partial t}$$
$$-\sqrt{\frac{1-\gamma R^*}{\varrho}}\sqrt{\frac{1-\gamma R^*}{\varrho}}\frac{\partial^2 \sigma}{\partial x^2}$$
$$\mp \frac{1}{\sqrt{\varrho}}\left(\sqrt{1-\gamma R^*}\frac{\partial}{\partial x}\sqrt{A+2\gamma B\sigma}\frac{\partial \sigma}{\partial t}\right.$$
$$\left. -\frac{\partial}{\partial t}\sqrt{A+2\gamma B\sigma}\sqrt{1-\gamma R^*}\frac{\partial \sigma}{\partial x}\right). \quad (3.2.5)$$

Obviously, it is sufficient to estimate the expression in parentheses on the right-hand side of (3.2.5). Note, however, that because of the identity

$$\frac{\partial}{\partial t}f(\sigma)\frac{\partial \sigma}{\partial x} \equiv \frac{\partial}{\partial x}f(\sigma)\frac{\partial \sigma}{\partial t},$$

where $f(\sigma)$ is an arbitrary smooth function (Sect. 1.3.2), and the commutativity of convolution operators with differentiation operators, the following identity holds:

$$\sqrt{1-\gamma R^*}\frac{\partial}{\partial x}\sqrt{A+2\gamma B\sigma}\frac{\partial \sigma}{\partial t} \equiv \sqrt{1-\gamma R^*}\frac{\partial}{\partial t}\sqrt{A+2\gamma B\sigma}\frac{\partial \sigma}{\partial x}$$
$$\equiv \frac{\partial}{\partial t}\sqrt{1-\gamma R^*}\sqrt{A+2\gamma B\sigma}\frac{\partial \sigma}{\partial x}.$$

Therefore,

$$\frac{1}{\sqrt{\varrho}}\left(\sqrt{1-\gamma R^*}\frac{\partial}{\partial x}\sqrt{A+2\gamma B\sigma}\frac{\partial \sigma}{\partial t}-\frac{\partial}{\partial t}\sqrt{A+2\gamma B\sigma}\sqrt{1-\gamma R^*}\frac{\partial \sigma}{\partial x}\right)$$
$$\equiv \frac{1}{\sqrt{\varrho}}\frac{\partial}{\partial t}\left(\sqrt{1-\gamma R^*}\sqrt{A+2\gamma B\sigma}-\sqrt{A+2\gamma B\sigma}\sqrt{1-\gamma R^*}\right)\frac{\partial \sigma}{\partial x},$$

whence, because of expansions

$$\sqrt{A+2\gamma B\sigma} = \sqrt{A}\left(1+\gamma \frac{B}{A}\sigma\right)-\dots,$$
$$\sqrt{1-\gamma R^*} = 1-\frac{\gamma}{2}R^*-\dots. \quad (3.2.6)$$

where dots denote terms of order $O(\gamma^2)$, we obtain

$$\frac{1}{\sqrt{\varrho}}\left(\sqrt{1-\gamma R^*}\frac{\partial}{\partial x}\sqrt{A+2\gamma B\sigma}\frac{\partial \sigma}{\partial t}-\frac{\partial}{\partial t}\sqrt{A+2\gamma B\sigma}\sqrt{1-\gamma R^*}\frac{\partial \sigma}{\partial x}\right)$$
$$\equiv \sqrt{\frac{A}{\varrho}}\frac{\partial}{\partial t}\left\{\left(1-\frac{\gamma}{2}R^*-\dots\right)\left(1+\gamma \frac{B}{A}\sigma-\dots\right)\right.$$
$$\left. -\left(1+\gamma \frac{B}{A}\sigma-\dots\right)\left(1-\frac{\gamma}{2}R^*-\dots\right)\right\}\frac{\partial \sigma}{\partial x}$$
$$\equiv -\frac{\gamma^2 B}{\sqrt{A\varrho}}\frac{\partial}{\partial t}(R^*\sigma-\sigma R^*)\frac{\partial \sigma}{\partial x}+\dots = O(\gamma^2). \quad (3.2.7)$$

Now the required result follows from (3.2.5,7).

Remark. It is clear that by using (3.2.6) the decomposition (3.2.4) can be simplified to

$$\left\{\frac{\partial}{\partial t}\sqrt{A}\left(1+\gamma\frac{B}{A}\sigma\right)\mp\frac{1}{\sqrt{\varrho}}\left(1-\frac{\gamma}{2}R^*\right)\frac{\partial}{\partial x}\right\}$$
$$\times\left\{\sqrt{A}\left(1+\gamma\frac{B}{A}\sigma\right)\frac{\partial}{\partial t}\pm\frac{1}{\sqrt{\varrho}}\left(1-\frac{\gamma}{2}R^*\right)\frac{\partial}{\partial x}\right\}\sigma=O(\gamma^2). \quad (3.2.8)$$

3.2.2 Why Can't the Factorization be Exact?

In (3.2.7) one can easily see the reason why factorization in Theorem 3.2.1 cannot be exact. This reason is the noncommutativity of operators of convolution and of multiplication by a function. It is also clear that the quantity $O(\gamma^2)$ on the right-hand side of (3.2.4) is not uniformly small for bounded σ. This quantity may turn out to be considerable in the domain of rapid change of the function σ (e.g., in the vicinity of wave fronts). However, domains where σ changes quickly generally have a small area, which, obviously, reduces their influence.

3.2.3 Single–Wave Equation

It is clear from (3.2.8) that our asymptotic approach gives the following equation which describes waves travelling to the right:

$$\sqrt{A\varrho}\left(1+\gamma\frac{B}{A}\sigma\right)\frac{\partial\sigma}{\partial t}+\left(1-\frac{\gamma}{2}R^*\right)\frac{\partial\sigma}{\partial x}=0. \quad (3.2.9)$$

Equation (3.2.9) has also been derived following other considerations [3.8,9]. This equation can be transformed in the following way. We multiply (3.2.9) by the operator

$$\left(1-\frac{\gamma}{2}R^*\right)^{-1}=1+\frac{\gamma}{2}R^*+\ldots.$$

Then it is evident that (3.2.9) coincides with

$$\sqrt{A\varrho}\left(1+\frac{\gamma}{2}R^*\right)\frac{\partial\sigma}{\partial t}+\gamma\sqrt{A\varrho}\frac{B}{A}\sigma\frac{\partial\sigma}{\partial t}+\frac{\partial\sigma}{\partial x}=0 \quad (3.2.10)$$

with accuracy to $O(\gamma^2)$.

It is easy to see that when the kernel $R(t)$ is close to a constant, (3.2.10) becomes, with an accuracy to infinitesimal quantities of higher order, a purely differential equation

$$\sqrt{A\varrho}\frac{\partial\sigma}{\partial t}+\frac{\gamma}{2}R(0)\sigma+\gamma\sqrt{A\varrho}\frac{B}{A}\sigma\frac{\partial\sigma}{\partial t}+\frac{\partial\sigma}{\partial x}=0$$

3.2.4 Condition on the Shock for the Stress Wave

We shall now derive the condition on the shock for stress waves satisfying the nonlinear wave equation (3.2.3) in the domain of smoothness.

We write the dynamic and kinematic compatibility conditions

$$[\sigma] = -\varrho[v]U ,$$
$$[v] = -[\varepsilon]U \qquad (3.2.11)$$

where U is the velocity of the wave front. It follows from these conditions that

$$[\sigma] = \varrho[\varepsilon]U^2 . \qquad (3.2.12)$$

On the other hand, from the constitutive equation (3.2.1) which we write in the equivalent form

$$\varepsilon - \gamma R^*\varepsilon = A\sigma + \gamma B\sigma^2 .$$

it follows that

$$[\varepsilon] = A[\sigma] + \gamma B[\sigma^2] , \qquad (3.2.13)$$

since the convolution $R^*\varepsilon$ is, obviously, a continuous function even if ε suffers a jump. Substituting (3.2.13) into (3.2.12) we obtain

$$U = \sqrt{\frac{[\sigma]}{\varrho(A[\sigma] + \gamma B[\sigma^2])}} . \qquad (3.2.14)$$

We again choose the positive value of the square root, which corresponds to the wave travelling to the right.

Taking into account that γ is small, we can transform (3.2.14) [with accuracy to $O(\gamma^2)$] into the form

$$U = \frac{1}{\sqrt{A\varrho}} \frac{1}{1 + \gamma(B/2A)([\sigma^2]/[\sigma])} . \qquad (3.2.15)$$

As in Chap. 1, letting in (3.2.15) $[\sigma] \to 0$, we obtain that weak shocks must propagate along characteristics.

Note now that the shock condition (3.2.15) has the same form as in the case of elasticity. Thus we can follow the arguments of Sect. 1.1.5 in obtaining the stability condition for the shock-wave front. This condition coincides with (1.1.17″) (where it is necessary to replace $a(\sigma)$ by $A\sigma + \gamma B\sigma^2$):

$$\begin{aligned}[\sigma] &> 0 \quad \text{for} \quad B < 0 , \\ [\sigma] &< 0 \quad \text{for} \quad B > 0 .\end{aligned} \qquad (3.2.16)$$

3.2.5 New Notation

Concluding Sect. 3.2, we somewhat simplify our notation by setting

$$\bar{x} = x\sqrt{A\varrho}, \quad k = -B/A.$$

Then (3.2.9,10), as one can easily check, will take the respective forms

$$\frac{\partial}{\partial t}\left(\sigma - \frac{k\gamma}{2}\sigma^2\right) + \left(1 - \frac{\gamma}{2}R^*\right)\frac{\partial \sigma}{\partial \bar{x}} = 0 \qquad (3.2.9')$$

and

$$\frac{\partial}{\partial t}\left(\sigma - \frac{k\gamma}{2}\sigma^2\right) + \frac{\partial \sigma}{\partial \bar{x}} + \frac{\gamma}{2}R^*\frac{\partial \sigma}{\partial t} = 0, \qquad (3.2.10')$$

and (3.2.15) becomes

$$\bar{U} = \frac{1}{1 - (k\gamma/2)([\sigma^2]/[\sigma])} \qquad (3.2.15')$$

(where $\bar{U}$ is the front velocity in the coordinates $t, \bar{x}$). Finally, since A is positive, the stability condition can be rewritten as

$$\begin{aligned}[\sigma] > 0 \quad &\text{for} \quad k > 0, \\ [\sigma] < 0 \quad &\text{for} \quad k < 0.\end{aligned} \qquad (3.2.16')$$

It is easy to show that the function σ, satisfying (3.2.9') in the domain of smoothness and the condition (3.2.15'), is a weak solution of (3.2.9') in the following sense: for any smooth test function $g = g(t, \bar{x})$ with a compact support

$$-\iint\left\{\left(\sigma - \frac{k\gamma}{2}\sigma^2\right)g'_t + \left(\sigma - \frac{\gamma}{2}R^*\sigma\right)g'_{\bar{x}}\right\}dt\,d\bar{x} = 0.$$

Conversely, a weak solution of (3.2.9'), which is continuous outside some smooth curve $t = t(\bar{x})$ and tends to finite one-sided limits when approaching this curve, satisfies the condition (3.2.15'). In the domain of smoothness this solution is obviously the solution of (3.2.9') in the classical sense.

The analogous assertions are also valid for (3.2.10').

Remark. In Sects. 3.3–3.6 we shall use (3.2.9'), (3.2.10') and (3.2.15') omitting the bar over U and x for simplicity.

3.3 Continuous Stationary Profile Waves and Nonzero Solutions of Homogeneous Integral Volterra Equations

3.3.1 Waves Propagating in an Undisturbed Medium

In this section the following question is of interest: do there exist, in the framework of the model (3.2.1), stress waves travelling in the undisturbed medium with no change in their form? To be exact, we consider only waves travelling to the right. In Sect. 3.3 we consider only waves which do not contain strong shocks; thus we deal with continuous (weak) solutions of (3.2.9'):

$$\frac{\partial}{\partial t}\left(\sigma - \frac{k\gamma}{2}\sigma^2\right) + \left(1 - \frac{\gamma}{2}R^*\right)\frac{\partial \sigma}{\partial x} = 0 \qquad (3.3.1)$$

(the bar over x is omitted).

Before we proceed further, we formulate all the necessary assumptions concerning (3.3.1).

Firstly, we surmise that $k > 0$. In the case of absence of memory, this condition corresponds to the overtaking of tension waves (travelling to the right) in the direction of wave motion. The case $k < 0$, which corresponds to the overtaking of compression waves in the direction of wave motion, can be reduced to the case $k > 0$ by the change $\sigma \to -\sigma$ in (3.3.1).

Secondly, we impose the following restrictions on the hereditary kernel $R(t)$:

a) $R(t) \geq 0$, $R'(t) \leq 0$ for $t > 0$,

b) $J \equiv \int_0^\infty R(t)dt < \infty$. $\qquad (3.3.2)$

3.3.2 Integral Equation for the Wave of Stationary Profile

We now return to the question posed in Sect. 3.3.1. It is clear that the question can be reformulated in the following way: does (3.3.1) have continuous solutions of the form

$$\sigma = f\left(t - \frac{x}{c}\right), \quad c > 0$$

where $f(z) = 0$ for $z \to -\infty$?

It is well-known that disturbances of infinitesimal amplitude propagate in an undisturbed medium with instantaneous elastic velocity[1] (in our coordinates the instantaneous elastic velocity, obviously, equals unity). However, it follows from the continuity of the function f that in the vicinity of the front the quantity σ will be infinitesimal. Hence, we have to seek the desired continuous solution of (3.3.1) in the form

---

[1] There are some interesting exceptions to this principle, due to the discontinuities which the coefficients of equations of motion may suffer [3.10-12].

$$\sigma = f(t - x) \,. \tag{3.3.3}$$

Substituting (3.3.3) into (3.3.1), we obtain

$$\left(f - \frac{k\gamma}{2}f^2\right)' - f' + \frac{\gamma}{2} R^* f' = 0 \,,$$

i.e.,

$$(k\gamma f^2)' - \gamma(R^* f)' = 0 \,,$$

whence

$$kf^2(t - x) - \int_{-\infty}^{t} R(t - \tau)f(\tau - x)d\tau = \text{const} \,.$$

Redenoting $z = t - x$ by t and taking into account the fact that const $= 0$ [since $f(t) = 0$ as $t \to -\infty$], we finally arrive at the equation

$$kf^2(t) - \int_{-\infty}^{t} R(t - \tau)f(\tau)d\tau = 0 \,. \tag{3.3.4}$$

3.3.3 Estimate of the Solution of the Integral Equation

It is clear that $f(t) = 0$ is a solution of (3.3.4). However, this solution is of no interest to us. Our purpose is to find nonzero solutions of (3.3.4) (which, however, become identically equal to zero as $t \to -\infty$).

Proposition. Let the function $f(t)$ be a solution of (3.3.4). Then for an arbitrary t_0 the function $f(t - t_0)$ is also a solution of (3.3.4).

Proof. Replacing t by $t - t_0$ in (3.3.4), we obtain

$$f^2(t - t_0) = \int_{-\infty}^{t-t_0} R(t - t_0 - \eta)f(\eta)d\eta$$

$$= \int_{-\infty}^{t} R(t - t_0 - (\tau - t_0))f(\tau - t_0)d\tau$$

$$= \int_{-\infty}^{t} R(t - \tau)f(\tau - t_0)d\tau \,,$$

which gives the required result.

Thus, if (3.3.4) has at least one nonzero solution (which becomes identically equal to zero as $t \to -\infty$), then it turns out that there is a whole family of such solutions. Evidently, in this family of solutions there must exist the solution which becomes identically equal to zero for $t \leq 0$ and differs from zero for however small positive t. This solution corresponds to the wave $\sigma = f(t - x)$ with the front passing through the origin of the Lagrangian coordinate $x = 0$

at the moment $t = 0$. We shall study this very solution (in supposition of its existence) in the following lemma.

Remark. Because the considered solution of (3.3.4) becomes identically equal to zero for $t \leq 0$, then, instead of (3.3.4), we can consider

$$kf^2(t) - \int_0^t R(t-\tau)f(\tau)d\tau = 0, \quad t \geq 0 \tag{3.3.5}$$

with the initial condition $f(0) = 0$ which follows from the continuity of $f(t)$ at zero. This initial condition appears superfluous, however, since it holds automatically for any reasonable solution of (3.3.5).

Lemma 3.3.3. Let there exist for (3.3.5) a bounded monotonically increasing solution $f(t)$, $t \geq 0$ which is not identically equal to zero on an arbitrarily small interval $(0, \delta)$. Then $f(t)$ is continuous for $t \geq 0$, $f(0) = 0$ and $f(t) > 0$ for $t > 0$. Moreover, the following relations are valid:

$$f(t) \to f(\infty) = \frac{J}{k} \quad \text{as} \quad t \to \infty \tag{3.3.6}$$

where $J \equiv \int_0^\infty R(t)dt$;

$$\frac{1}{2k}\sqrt{R(t)}\int_0^t \sqrt{R(t)}dt \leq f(t) \leq \frac{1}{k}\int_0^t R(t)dt. \tag{3.3.7}$$

Proof. Let $f(t)$, $t \geq 0$ be a solution of (3.3.5) with the properties inidicated in the condition of the lemma. Then the integral term in (3.3.5) is obviously a continuous function for $t \geq 0$ and vanishes for $t = 0$. Therefore, $f(t)$ is also a continuous function for $t \geq 0$ vanishing for $t = 0$. The fact that $f(t) > 0$ for $t > 0$ is now obvious from the condition of the lemma.

If the kernel $R(t)$ is infinitely differentiable for $t > 0$ and

$$|R^{(n)}(0)| < \infty; \quad n = 0, 1, \ldots,$$

then $f(t)$ also proves to be infintely differentiable for $t > 0$. The validity of the fact is obvious from (3.3.5).

Letting t tend to infinity in (3.3.5), if follows that

$$kf^2(\infty) = f(\infty)\int_0^\infty R(\tau)d\tau.$$

Therefore, (3.3.6) follows.

Furthermore, using the monotonic increase of $f(t)$ and its established non-negativeness, we obtain from (3.3.5)

$$kf^2(t) \leq f(t)\int_0^t R(t)dt$$

which leads to the right inequality in (3.3.7).

To prove the left inequality in (3.3.7), we must note that (3.3.5) yields

$$kf^2(t) \geq R(t) \int_0^t f(t)dt$$

since $R(t)$ is a nonnegative function. Let us denote

$$j(t) = \int_0^t f(t)dt \,;$$

then the preceding inequality will become

$$k(j'(t))^2 \geq R(t)j(t)$$

whence [because $j(t)$ is not identically equal to zero!]

$$\frac{j'(t)}{\sqrt{j(t)}} \geq \left(\frac{1}{k}\right)^{1/2} \sqrt{R(t)} \,. \tag{3.3.8}$$

Integrating this inequality and taking into account that $j(0) = 0$, we obtain

$$2\sqrt{j(t)} \geq \left(\frac{1}{k}\right)^{1/2} \int_0^t \sqrt{R(t)}dt \,. \tag{3.3.8'}$$

The required inequality follows from (3.3.8) and (3.3.8'):

$$j'(t) \geq \frac{1}{2}\left(\frac{1}{k}\right)^{1/2}\left(\frac{1}{k}\right)^{1/2} \sqrt{R(t)} \int_0^t \sqrt{R(t)}dt \,.$$

The lemma is proved.

3.3.4 Existence of Stationary Profile Waves. Special Case

Below we see that the solution $f(t)$ which is discussed in Lemma 3.3.3 in fact exists. However, before we start proving this purely mathematical result, we shall consider a special case where the kernel $R(t)$ has a particular analytical form for small $t > 0$.

Suppose

$$R(t) = Ct^{\alpha-1} \quad \text{for} \quad 0 < t < t_0 \tag{3.3.9}$$

where $0 < \alpha \leq 1$, $C > 0$. We seek the solution of (3.3.5) for $0 \leq t < t_0$ in the form

$$f(t) = Dt^\beta \,.$$

Inserting this expression into (3.3.5) [with the kernel (3.3.9)], we obtain for $0 \leq t < t_0$

$$kD^2t^{2\beta} - CD\int_0^t (t-\tau)^{\alpha-1}\tau^{(\beta+1)-1}d\tau = 0,$$

i.e.,

$$kDt^{2\beta} - C\frac{\Gamma(\alpha)\Gamma(\beta+1)}{\Gamma(\alpha+\beta+1)}t^{\alpha+\beta} = 0$$

whence we have immediately the following:

$$\alpha = \beta, \quad D = C\frac{\Gamma(\alpha)\Gamma(\alpha+1)}{k\Gamma(2\alpha+1)}$$

where Γ is the Euler's Gamma function. Thus, the function

$$f(t) = \frac{C\Gamma(\alpha)\Gamma(\alpha+1)}{k\Gamma(2\alpha+1)}t^\alpha \qquad (3.3.10)$$

is an exact solution of (3.3.5,9) on the interval $[0, t_0)$.

Theorem 3.3.4. (preparatory) [3.13]. Let the kernel $R(t)$ satisfy the condition (3.3.9) as well as the general conditions (3.3.2). Then for (3.3.5) there exists a bounded monotonically increasing solution $f(t)$ defined on the entire semi-axis $t \geq 0$ and coincident with (3.3.10) on $[0, t_0)$. (For this solution all the assertions of Lemma 3.3.3 are obviously valid.)

Proof. We use the method of successive approximations. Since the desired nonzero solution of (3.3.5) cannot be unique a fortiori, it is important to correctly choose the zero approximation to obtain convergent successive approximations.

Let us set

$$f_0(t) = \frac{C\Gamma(\alpha)\Gamma(\alpha+1)}{k\Gamma(2\alpha+1)}t^\alpha \quad \text{for} \quad 0 \leq t < t_0,$$
$$f_0(t) = \frac{J}{k} \quad \text{for} \quad t \geq t_0 \qquad (3.3.11)$$

(it is clear that $f_0(t)$ is a nondecreasing function), and let us define the successive approximations recurrently in the following way:

$$kf_1^2(t) = \int_0^t R(t-\tau)f_0(\tau)d\tau, \qquad (3.3.12)$$

$$kf_2^2(t) = \int_0^t R(t-\tau)f_1(\tau)d\tau, \qquad (3.3.13)$$

...

For $f_n(t)$, $n = 1, 2, \ldots$, we take the positive value of the square root of the right-hand side of the corresponding equality, i.e.,

$$f_n = \frac{1}{\sqrt{k}} \sqrt{\int_0^t R(t-\tau) f_{n-1}(\tau) d\tau} \; .$$

It is clear that

$$f_1(t) \equiv f_0(t) \quad \text{for} \quad 0 \le t < t_0 \; . \tag{3.3.14}$$

Moreover, since $R(t)$ is monotonically decreasing and $f_0(t)$ is monotonically increasing, it follows from (3.3.12) that $f_1(t)$ is monotonically increasing. Finally, note that letting $t \to \infty$ in (3.3.12) we obtain

$$k f_1^2(\infty) = f_0(\infty) \int_0^\infty R(t) dt = f_0(\infty) J$$

whence because of (3.3.11)

$$k f_1^2(\infty) = \frac{J}{k} J \; ,$$

and, therefore

$$f_1(\infty) = \frac{J}{k} = f_0(\infty) \; . \tag{3.3.15}$$

Thus we can easily see that $0 \le f_1(t) \le f_0(t)$ on the entire semi-axis $t \ge 0$ (Fig. 3.1).

Now let us consider the difference between (3.3.13) and (3.3.12). We have

$$k \left(f_2^2(t) - f_1^2(t) \right) = \int_0^t R(t-\tau) \left(f_1(\tau) - f_0(\tau) \right) d\tau \le 0 \; ,$$

i.e.,

$$k \left(f_2(t) + f_1(t) \right) \left(f_2(t) - f_1(t) \right) \le 0$$

whence, since f_1 and f_2 are nonnegative, it follows that

$$0 \le f_2(t) \le f_1(t) \quad \text{for} \quad t \ge 0 \; .$$

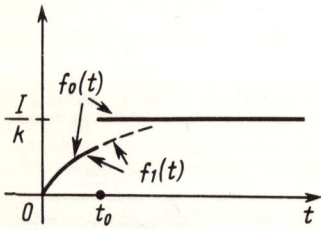

Fig. 3.1. The functions $f_0(t)$ and $f_1(t)$ are shown

Besides, from (3.3.13,15) one can easily see that

$$f_2(\infty) = f_0(\infty) = J/k$$

and from (3.3.13,14) it follows that

$$f_2(t) \equiv f_0(t) \quad \text{for} \quad 0 \leq t < t_0 .$$

Similarly, by induction we obtain

$$0 \leq f_n(t) \leq f_{n-1}(t) \leq \ldots \leq f_0(t) \quad \text{for} \quad t \geq 0 \qquad (3.3.16)$$

where f_i, $i = 1, 2, \ldots$ are monotonically increasing continuous functions tending to the limit J/k as $t \to \infty$ and such that

$$f_i(t) \equiv f_0(t) \quad \text{for} \quad 0 \leq t \leq t_0 . \qquad (3.3.17)$$

Therefore, there exists the nonzero limiting function

$$f(t) = \lim_{n \to \infty} f_n(t) \qquad (3.3.18)$$

which is the desired solution of (3.3.5,9). It is clear that $f(t)$ is an increasing bounded function having the form of (3.3.10) for $0 \leq t < t_0$. The theorem is proved.

3.3.5 Existence of the Wave of Stationary Profile. General Case

Let $R(t)$ now be an arbitrary function satisfying only the general geometric restrictions (3.3.2). Let us define the sequence of kernels

$$R_n(t) = R\left(\frac{1}{n}\right) \quad \text{for} \quad 0 \leq t < \frac{1}{n} ,$$
$$R_n(t) = R(t) \quad \text{for} \quad t \geq \frac{1}{n} \qquad (3.3.19)$$

($n = 1, 2, \ldots$). It is clear that

$$R_1(t) \leq R_2(t) \leq \ldots \leq R_n(t) \leq \ldots \leq R(t) . \qquad (3.3.20)$$

Moreover, all the $R_n(t)$ are decreasing functions and each of them satisfies the condition (3.3.9) with $C = R(1/n)$, $\alpha = 1$, $t_0 = 1/n$, respectively. So, using Theorem 3.3.4 and Lemma 3.3.3, every equation

$$kf^2(t,n) = \int_0^t R_n(t-\tau)f(\tau,n)d\tau , \quad \text{with} \quad n = 0, 1, 2, \ldots \qquad (3.3.21)$$

has a continuous monotonically increasing bounded solution $f(t,n)$ such that $f(0,n) = 0$ and

$$f(t,n) \longrightarrow \frac{\int_0^\infty R_n(\tau)d\tau}{k} \leq \frac{\int_0^\infty R(\tau)d\tau}{k} , \quad t \to \infty . \qquad (3.3.22)$$

Furthermore, according to Theorem 3.3.4, each of the functions $f(t,n)$, $n = 0, 1, \ldots$ is the limit to the successive approximations defined by formulas of the type (3.3.11–13). Namely,

$$f(t,n) = \lim_{i \to \infty} f_i(t,n)$$

where

$$f_0(t,n) = R\left(\frac{1}{n}\right) \frac{\Gamma(1)\Gamma(2)}{k\Gamma(3)} t \quad \text{for} \quad 0 \leq t < \frac{1}{n},$$

$$f_0(t,n) = \frac{1}{k} \int_0^\infty R_n(\tau) d\tau \quad \text{for} \quad t > \frac{1}{n};$$

$$k f_1^2(t,n) = \int_0^t R_n(t-\tau) f_0(\tau, n) d\tau,$$

$$k f_2^2(t,n) = \int_0^t R_n(t-\tau) f_1(\tau, n) d\tau, \qquad (3.3.23)$$

$\ldots$

But from (3.3.23), by virtue of (3.3.20) and the obvious fact that

$$R(1) < R\left(\frac{1}{2}\right) < \ldots < R\left(\frac{1}{n}\right) \ldots,$$

it easily follows that for $t > 0$

$$0 < f_0(t,1) \leq f_0(t,2) \leq \ldots \leq f_0(t,n) \leq \ldots,$$

$$0 < f_1(t,1) \leq f_1(t,2) \leq \ldots \leq f_1(t,n) \leq \ldots, \qquad (3.3.24)$$

$\ldots$

Therefore, the following analogous inequalities are evidently valid for the limiting functions:

$$0 < f(t,1) \leq f(t,2) \leq \ldots \leq f(t,n) \leq \ldots. \qquad (3.3.25)$$

But, from a monotonic increase of the nonnegative functions $f(t,n)$ and from the relation (3.3.22), it follows that all of these functions are bounded by one and the same constant:

$$f(t,n) \leq \frac{1}{k} \int_0^\infty R(\tau) d\tau. \qquad (3.3.26)$$

Now (3.3.25,26) and monotonicity of the functions $f(t,n)$ yield the existence of the limit

$$f(t) = \lim_{n \to \infty} f(t,n).$$

Therefore in (3.3.21) we can pass to the limit as $n \to \infty$, whence we have the following result.

Theorem 3.3.5 (basic) [3.13]. Let the kernel $R(t)$ satisfy the geometric restrictions (3.3.2). Then the equation

$$kf^2(t) - \int_0^t R(t-\tau)f(\tau)d\tau = 0, \quad t \geq 0$$

has a nonnegative bounded monotonically increasing solution which is distinct from zero for however small $t > 0$. All the assertions of Lemma 3.3.3 are obviously valid for this solution.

Remark. Similarly, one can construct a nonzero solution of a more general equation

$$F(f) - \int_0^t R(t-\tau)f(\tau)d\tau = 0, \quad t \geq 0 \qquad (3.3.27)$$

where $F(f)$ is a monotonically increasing function which can be represented as $c_2 f^2 + c_3 f^3 + \ldots$.

3.3.6 The Exponential Kernel

Let us take, as an example, the case of

$$R(t) = Ce^{-\alpha t}, \quad t \geq 0 \qquad (3.3.28)$$

where $\alpha > 0$, $C > 0$ [3.8,14]. Then the equation (3.3.5) will take the form

$$f^2(t) = C_1 \int_0^t e^{-\alpha(t-\tau)} f(\tau)d\tau, \quad C_1 = \frac{C}{k}. \qquad (3.3.29)$$

Differentiating (3.3.29) with respect to t, we obtain

$$2ff' = C_1 f - C_1 \alpha \int_0^t e^{-\alpha(t-\tau)} f(\tau)d\tau. \qquad (3.3.30)$$

By eliminating the integral term from (3.3.29,30), we arrive at the differential equation

$$\alpha f^2 + 2ff' = C_1 f$$

whence, by cancellation by f (as we are seeking a nonzero solution), we obtain

$$f' + \frac{\alpha}{2} f = \frac{C_1}{2}. \qquad (3.3.31)$$

The general solution of (3.3.31) has the form

$$f(t) = \frac{C_1}{\alpha} + C_2 e^{-\alpha t/2}.$$

However, we are not interested in just any solution, but in one that becomes zero for $t = 0$. Hence, $C_2 = -C_1/\alpha$ and

$$f(t) = \frac{C_1}{\alpha}\left(1 - e^{-\alpha t/2}\right), \quad t \geq 0. \tag{3.3.32}$$

It is easy to check that this function is in fact a solution of (3.3.29).

Problem. Prove that the solution of (3.3.5), which equals zero for $t \leq 0$ and differs from zero for however small $t > 0$, is unique. It is assumed that the conditions of (3.3.2) hold.

3.3.7 The Simplest Oscillatory Kernel

In the previous section the monotonicity and nonnegativeness of the kernel $R(t)$ yielded the existence, monotonicity and nonnegativeness of the nonzero solution $f(t)$ of (3.3.5). However, as mentioned, there are models in which oscillatory kernels of memory are used. For (3.3.5) to have a nonzero solution it is apparently enough that the kernel $R(t)$ is positive for small $t > 0$.

Let us consider the simplest oscillatory kernel

$$R(t) = C \sin \omega t, \quad t \geq 0 \tag{3.3.33}$$

where $\omega > 0$, $C > 0$. Then (3.3.5) will take the form

$$f^2(t) = C_1 \int_0^t f(\tau) \sin \omega(t - \tau) d\tau,$$
$$C_1 = \frac{C}{k}. \tag{3.3.34}$$

Successively differentiating this equation, we obtain

$$(f^2)' = C_1 \omega \int_0^t f(\tau) \cos \omega(t - \tau) d\tau, \tag{3.3.35}$$

$$(f^2)'' = C_1 \omega f(t) - C_1 \omega^2 \int_0^t f(\tau) \sin \omega(t - \tau) d\tau. \tag{3.3.36}$$

Elimination of the integral summand from (3.3.34,36) yields an ordinary differential equation

$$(f^2)'' = C_1 \omega f - \omega^2 f^2. \tag{3.3.37}$$

Multiplying both parts of this equation by $(f^2)' = 2ff'$, we obtain

$$d\frac{1}{2}\left((f^2)'\right)^2 = d\left(\frac{2}{3} C_1 \omega f^3 - \frac{\omega^2}{2} f^4\right)$$

whence

$$\frac{1}{\sqrt{2}}(f^2)' = \sqrt{\frac{2}{3}C_1\omega f^3 - \frac{\omega^2}{2}f^4} + \text{const} .$$

However, it evidently follows from (3.3.35) that $(f^2)' = 0$ for $t = 0$. Taking into account this fact and that $f(0) = 0$, we obtain that in the previous equality we must have const = 0. Therefore the previous equality yields

$$\frac{1}{\sqrt{2}}\int_0^f \frac{df^2}{\sqrt{(2/3)C_1\omega f^3 - (\omega^2/2)f^4}} = t + \text{const} .$$

But here also const = 0 since $f(0) = 0$. Thus, we arrive at the equality

$$2\frac{1}{\sqrt{2}}\int_0^f \frac{df}{\sqrt{(2/3)C_1\omega f - (1/2)\omega^2 f^2}} = t$$

whence

$$f(t) = \frac{2C_1}{3\omega}\left(1 - \cos\frac{\omega t}{2}\right) . \tag{3.3.38}$$

By direct substitution it is easy to make sure that the function (3.3.38) really satisfies the equation (3.3.34).

3.3.8 A More Complicated Oscillatory Kernel

We now consider a more general case of

$$R(t) = Ce^{-\alpha t}\sin(\omega t + \varphi_0) , \quad t \geq 0 \tag{3.3.39}$$

where $C > 0$, $\alpha \geq 0$, $\omega > 0$, $0 \leq \varphi_0 < \pi$. Then the integral equation (3.3.5) takes the form

$$f^2(t) = C_1 \int_0^t f(\tau)e^{-\alpha(t-\tau)}\sin(\omega(t-\tau) + \varphi_0)\,d\tau , \tag{3.3.40}$$

$$C_1 = C/k .$$

Differentiating (3.3.40) with respect to t (and reducing similar terms), we successively obtain

$$(f^2)' = C_1\left\{f(t)\sin\varphi_0 - \alpha\int_0^t f(\tau)e^{-\alpha(t-\tau)}\sin(\omega(t-\tau) + \varphi_0)\,d\tau \right.$$
$$\left. + \omega\int_0^t f(\tau)e^{-\alpha(t-\tau)}\cos(\omega(t-\tau) + \varphi_0)\,d\tau\right\} ,$$

3. Nonlinear Waves in Media with Memory

$$(f^2)'' = C_1 \bigg\{ f'(t) \sin \varphi_0 - f(t) \alpha \sin \varphi_0$$
$$+ (\alpha^2 - \omega^2) \int_0^t f(\tau) e^{-\alpha(t-\tau)} \sin(\omega(t-\tau) + \varphi_0) \, d\tau$$
$$+ \omega f(t) \cos \varphi_0 - 2\alpha\omega \int_0^t f(\tau) e^{-\alpha(t-\tau)} \cos(\omega(t-\tau) + \varphi_0) \, d\tau \bigg\} \quad (3.3.41)$$

Eliminating the integral terms form (3.3.40,41), we arrive at the ordinary nonlinear differential equation of the second order

$$(f^2)'' + 2\alpha(f^2)' - C_1 f' \sin \varphi_0 - C_1 (\omega \cos \varphi_0 + \alpha \sin \varphi_0) f$$
$$+ (\alpha^2 + \omega^2) f^2 = 0 \,. \quad (3.3.42)$$

It is easy to determine the initial conditions for (3.3.42). First of all, $f(0) = 0$ by virtue of (3.3.40). To determine the value of $f'(0)$, we have to pass to the limit as $t \to +0$ in the first of the equalities (3.3.41). As a result we obtain

$$2f(t)f'(t) = C_1 \{ f(t) \sin \varphi_0 - \alpha f(t) o(1) + \omega f(t) o(1) \} \,, \quad t \to +0$$

whence

$$f'(0) = \frac{C_1 \sin \varphi_0}{2} \,.$$

Unfortunately, in the general case it is impossible to find the analytic solution of (3.3.42).

Problem. a) Prove that in the case, where

$$\int_0^\infty R(t) dt = 0$$

and $R(t) > 0$ for small $t > 0$, the equation (3.3.5) has a nonzero solution vanishing as $t \to \pm\infty$.

b) Prove that if

$$\int_0^t R(t) dt > 0$$

for all finite $t > 0$, then the solution mentioned in a) is nonnegative.

3.3.9 Waves Propagating in a Pre-Stressed Medium

There is no doubt that waves of stationary profile when the entire medium is disturbed and it is impossible to mark out the wave front in front of which stress is equal to zero, are also of interest.

In this case we are not able to give a rigorous mathematical demonstration of the conclusions we arrive at intuitively, because the nonlinear integral equation that appears here does not admit a simple geometric approach. However, to support our conclusions we shall consider, as an example, the case of the exponential hereditary kernel for which the solution can be found analytically.

Therefore, we shall seek the continuous solution of the single-wave equation (3.3.1)

$$\frac{\partial}{\partial t}\left(\sigma - \frac{k\gamma}{2}\sigma^2\right) + \left(1 - \frac{\gamma}{2}R^*\right)\frac{\partial \sigma}{\partial x} = 0$$

in the form

$$\sigma = f\left(t - \frac{x}{c}\right), \quad c > 0.$$

The direct substitution yields

$$\left(f - \frac{k\gamma}{2}f^2\right)' - \frac{1}{c}f' + \frac{\gamma}{2c}R^*f' = 0.$$

It is clear that $f = \text{const}$ is always the solution of this equation; however, we are interested only in solutions which are not identically constant. Integrating the previous equation and redenoting $z = t - x/c$ by t, we obtain

$$\left(1 - \frac{1}{c}\right)f(t) - \frac{k\gamma}{2}f^2(t) + \frac{\gamma}{2c}\int_{-\infty}^{t}R(t-\tau)f(\tau)d\tau = \tilde{C}_0, \quad \tilde{C}_0 = \text{const}.$$

This time, however, we cannot consider

$$\tilde{C}_0 = 0;$$

moreover, the velocity of the wave c also remains unknown. For convenience we write the obtained equation in the equivalent form as follows:

$$f^2(t) - \frac{2}{k\gamma}\left(1 - \frac{1}{c}\right)f(t) - \frac{1}{kc}\int_{-\infty}^{t}R(t-\tau)f(\tau)d\tau = C_0 \qquad (3.3.43)$$

where

$$C_0 = -\frac{2}{k\gamma}\tilde{C}_0.$$

Suppose now that the solution of (3.3.43) tends to finite limits as $t \to \pm\infty$. We denote these limits by $f(\infty)$ and $f(-\infty)$, respectively. Then, in (3.3.43), taking the limit as $t \to \infty$, we easily obtain

3. Nonlinear Waves in Media with Memory

$$f^2(\infty) - \frac{2}{k\gamma}\left(1 - \frac{1}{c}\right) f(\infty) - \frac{J}{kc} f(\infty) = C_0 . \tag{3.3.44}$$

Here we make use of the fact that

$$J \equiv \int_0^\infty R(\tau) d\tau < \infty .$$

On the other hand, in (3.3.43), taking the limit as $t \to -\infty$, we easily obtain the analogous equality

$$f^2(-\infty) - \frac{2}{k\gamma}\left(1 - \frac{1}{c}\right) f(-\infty) - \frac{J}{kc} f(-\infty) = C_0 . \tag{3.3.44'}$$

Thus, the quantities $f(\infty)$ and $f(-\infty)$ appear to be the roots of the same quadratic equation

$$P(f) \equiv f^2 - \left(\frac{2}{k\gamma}\left(1 - \frac{1}{c}\right) + \frac{J}{kc}\right) f - C_0 = 0 . \tag{3.3.45}$$

Furthermore, by virtue of our supposition that in (3.3.1) $k > 0$, it is intuitively clear that for the wave of steady profile $\sigma = f(t - x/c)$, $f \neq$ const, the inequality

$$f(\infty) > f(-\infty) \tag{3.3.46}$$

must hold. In this case nonlinearity tends to cause greater stress values to move with greater speeds, while relaxation maintains the form of the wave profile.

So, real quantities $f(\infty)$ and $f(-\infty)$ turn out to be distinct. Therefore, the discriminant of (3.3.45) must be positive:

$$\left(\frac{1}{k\gamma}\left(1 - \frac{1}{c}\right) + \frac{J}{2kc}\right)^2 + C_0 > 0 .$$

Now equating the left-hand sides of (3.3.44) and (3.3.44'), we have

$$f^2(\infty) - \left(\frac{2}{k\gamma}\left(1 - \frac{1}{c}\right) + \frac{J}{kc}\right) f(\infty)$$

$$= f^2(-\infty) - \left(\frac{2}{k\gamma}\left(1 - \frac{1}{c}\right) + \frac{J}{kc}\right) f(-\infty) ,$$

i.e.,

$$f^2(\infty) - f^2(-\infty) = \left(\frac{2}{k\gamma}\left(1 - \frac{1}{c}\right) + \frac{J}{kc}\right) (f(\infty) - f(-\infty))$$

whence, by cancellation by $(f(\infty) - f(-\infty))$, we obtain the following expression for wave velocity

$$c = \frac{1 - \gamma J/2}{1 - (f(\infty) + f(-\infty))k\gamma/2} . \tag{3.3.47}$$

In particular, for $f(-\infty) = 0$ we have

$$c = \frac{1 - \gamma J/2}{1 - f(\infty)k\gamma/2}. \qquad (3.3.47')$$

This expression will also be obtained in another way in Sect. 3.5.

It is clear, however, that even with (3.3.46) being valid, a continuous wave of steady profile can exist but not for any two real values $f(\infty)$ and $f(-\infty)$ as the sufficiently great difference between these two values implies formation of the shock. In Sect. 3.3.10 we shall consider this problem in the particular case of the exponential function of memory. In this case we shall also give a rigorous proof of existence of the limits $f(\infty)$ and $f(-\infty)$ for the wave of stationary profile, provided its amplitude is bounded. Hence, (3.3.47) for the wave velocity will be rigorously proved for the case of an exponential kernel.

Remark. The methods of studying continuous stationary waves travelling in the undisturbed medium can be applied with the same success to the study of continuous stationary waves travelling in the homogeneously prestressed medium. We leave it to the reader to make the appropriate calculations.

3.3.10 The Exponential Kernel

Let

$$R(t) = Ae^{-\alpha t}, \quad t \geq 0 \qquad (3.3.48)$$

where $A > 0$, $\alpha > 0$. We shall seek the solution of the integral equation (3.3.43) [with the kernel (3.3.48)] in the class of bounded smooth functions defined on the whole t-axis. For simplicity, we do not consider weak shocks.

Differentiating (3.3.43) with respect to t and eliminating the integral summand in a standard way, we arrive at the following ordinary differential equation (in this connection see also [3.8]):

$$\left(\frac{1}{k\gamma}\left(1 - \frac{1}{c}\right) - f\right)\frac{df}{dt} = \frac{\alpha}{2}P(f) \qquad (3.3.49)$$

where

$$P(f) \equiv f^2 - \left(\frac{2}{k\gamma}\left(1 - \frac{1}{c}\right) + \frac{A}{kc\alpha}\right)f - C_0. \qquad (3.3.50)$$

Note that since the integral of the kernel (3.3.48) equals A/α, the quadratic trinomial (3.3.50) obviously coincides with the quadratic trinomial (3.3.45) introduced above for the kernel of a general form.

It is clear that $f \equiv (1 - 1/c)/k\gamma$ is a solution of (3.3.49) if $(1 - 1/c)/k\gamma$ is a root of the quadratic equation $P(f) = 0$. However, we have already agreed to ignore solutions which are identically equal to a constant. The solutions of

(3.3.49) which are not identically constant can be evidently represented in the form

$$\int \frac{\frac{1-1/c}{k\gamma} - f}{P(f)} \, df = \frac{\alpha}{2} t + \text{const} \tag{3.3.51}$$

where the indeterminate constant is of no significance to us as it corresponds only to a shift of the solution along the t-axis.

We now note that if the quadratic equation $P(f) = 0$ has no real roots, then for bounded f the left-hand side of (3.3.51) is always a bounded quantity. But this implies that (3.3.51) cannot define a bounded function determined on the whole t-axis. Then, it is easy to show by immediate integration that (3.3.51) cannot define a bounded function which is determined on the whole t-axis, when the quadratic equation $P(f) = 0$ has two coincident roots. Hence, in order for (3.3.51) to define a bounded function $f(t)$, $-\infty < t < \infty$, it is necessary that the roots of the quadratic equation $P(f) = 0$ should be real and distinct. This condition is familiar to us, as it was derived in Sect. 3.3.9 (though not rigorously) in a more general case and other considerations were regarded.

Therefore, we suppose that the roots of the quadratic equation $P(f) = 0$ are real and distinct. This time we denote them by f_1 and f_2; to be exact, we suppose that

$$f_2 > f_1 .$$

It is clear that any primitive function of the integrand A of (3.3.51)

$$A = \frac{\frac{1-1/c}{k\gamma} - f}{P(f)}$$

comprises three smooth branches defined on $(-\infty, f_1)$, (f_1, f_2) and (f_2, ∞), respectively. It is also evident that the solution of interest to (3.3.49) can be defined only by the middle branch, i.e.,

$$\int_\xi^f \frac{\frac{1-1/c}{k\gamma} - f}{P(f)} \, df = \frac{\alpha}{2} t + \text{const} \tag{3.3.51'}$$

where $\xi \in (f_1, f_2)$. For (3.3.51) to actually define a bounded smooth function $f = f(t)$, $-\infty < t < \infty$, it is necessary that the derivative with respect to f of the left-hand side of (3.3.51') should retain the sign on the interval (f_1, f_2), i.e., we must have

$$\frac{1-1/c}{k\gamma} \leq f_1 \quad \text{or} \quad \frac{1-1/c}{k\gamma} \geq f_2 .$$

One can easily show that the case where $(1 - 1/c)/k\gamma$ coincides with one of the roots, f_1 or f_2, must be excluded, because in this case the function $f = f(t)$, defined by (3.3.51'), turns out to be nonbounded.

Thus, one of the following two systems of inequalities must hold:

$$\begin{cases} f_1 < f_2, \\ \dfrac{1-1/c}{k\gamma} < f_1 \end{cases} \quad \text{or} \quad \begin{cases} f_1 < f_2, \\ \dfrac{1-1/c}{k\gamma} > f_2. \end{cases} \qquad (3.3.52)$$

The quantity c can obviously be expressed by f_1 and f_2 by means of a formula similar to (3.3.47):

$$c = \frac{1 - \frac{\gamma A}{2\alpha}}{1 - \frac{k\gamma}{2}(f_1 + f_2)}$$

therefore (3.3.52) can be rewritten as

$$\begin{cases} f_1 < f_2, \\ 1 - \dfrac{1 - \frac{k\gamma}{2}(f_1 + f_2)}{1 - \frac{\gamma A}{2\alpha}} < k\gamma f_1 \end{cases}$$

or

$$\begin{cases} f_1 < f_2, \\ 1 - \dfrac{1 - \frac{k\gamma}{2}(f_1 + f_2)}{1 - \frac{\gamma A}{2\alpha}} > k\gamma f_2. \end{cases}$$

It is easy to check that the latter system is inconsistent, while the former one is equivalent to the following inequalities:

$$f_1 < f_2 < \left(1 - \frac{\gamma A}{\alpha}\right) f_1 + \frac{A}{\alpha k}. \qquad (3.3.53)$$

Now it is clear that provided the inequalities of (3.3.53) hold, the left-hand side of (3.3.51') is a monotonically increasing function of f tending to $-\infty$ as $f \to f_1$ and ∞ as $f \to f_2$. Thus

$$f_1 = f(-\infty), \quad f_2 = f(\infty). \qquad (3.3.54)$$

In consideration of this and the fact that

$$\frac{A}{\alpha} = J \equiv \int_0^\infty R(\tau) d\tau,$$

the condition (3.3.53) will finally take the form

$$f(-\infty) < f(\infty) < (1 - \gamma J) f(-\infty) + \frac{J}{k}. \qquad (3.3.55)$$

This is the desired condition of smoothness for the wave of stationary profile.

Notice that for $f(-\infty) = 0$ the condition (3.3.55) assumes the form

$$0 < f(\infty) < \frac{J}{k},$$

which is obviously in accordance with (3.3.6) where for the wave propagating in the undisturbed medium and having a weak shock on the front we have

$$f(\infty) = \frac{J}{k}.$$

Finally, integrating (3.3.51) and taking into account (3.3.54) and the equality (3.3.47) for the wave velocity, we easily obtain the following analytical formula for the smooth wave of stationary profile:

$$(f - f(-\infty))^{\frac{f(-\infty)-B}{f(\infty)-f(-\infty)}} (f(\infty) - f)^{-\frac{f(\infty)-B}{f(\infty)-f(-\infty)}} = \text{const } e^{\alpha t/2} \qquad (3.3.56)$$

where

$$B = \frac{1}{2} \frac{f(-\infty) + f(+\infty) - J/k}{1 - \gamma J/2}, \quad \text{const} > 0. \qquad (3.3.57)$$

3.4 Stationary Profile Shock-Waves and Self-Coordinated Integral Volterra Equations

3.4.1 Waves Propagating in an Undisturbed Medium

We again begin with the study of stress waves travelling to the right in the undisturbed medium without change of the form:

$$\sigma = g(t - x/c), \quad c > 0. \qquad (3.4.1)$$

Now we are interested in the case where waves have strong shocks on the front. As in Sect. 3.3, for convenience we consider σ as the generalized solution of the equation

$$\frac{\partial}{\partial t}\left(\sigma - \frac{k\gamma}{2}\sigma^2\right) + \left(1 - \frac{\gamma}{2}R^*\right)\frac{\partial \sigma}{\partial x} = 0, \quad k > 0, \quad 0 < \gamma \ll 1, \qquad (3.4.2)$$

which implies that the shock condition (3.2.15') holds:

$$U = \left(1 - \frac{k\gamma}{2}\frac{[\sigma^2]}{[\sigma]}\right)^{-1}. \qquad (3.4.3)$$

Because of the agreement made in Sect. 3.2.5, we omit the bar over the quantities U and x.

We again suppose that the kernel of memory $R(t)$ satisfies the conditions

$$R(t) \geq 0, \quad R'(t) \leq 0 \quad \text{for} \quad t > 0,$$

$$J \equiv \int_0^\infty R(t)dt < \infty. \qquad (3.4.4)$$

3.4.2 Integral Equation for Stationary Profile Waves

Without loss of generality we consider the wave front to pass through the origin of the system of coordinates at the moment $t = 0$. This corresponds to the supposition that in (3.4.1)

$$g(z) = 0 \quad \text{for} \quad z < 0 \tag{3.4.5}$$

and $g(+0) \neq 0$. For convenience we extend the definition of the function g to zero by right continuity

$$g(0) = g(+0) \,.$$

As a result, we are able to write $g(0)$ instead of $g(+0)$.

It is clear that in (3.4.1) c is the velocity of the shock-wave front and $g(0)$ is the value of the stress jump on this front. Thus, by virtue of (3.4.3) c and $g(0)$ turn out to be linked by the relation

$$c = \frac{1}{1 - \frac{k\gamma}{2} g(0)} \,. \tag{3.4.6}$$

Then, since we suppose that $k > 0$, the stability condition (3.2.16') takes the form

$$g(0) > 0 \,. \tag{3.4.7}$$

Therefore, it follows from (3.4.6) that $c > 1$.

Let us now insert (3.4.1) into (3.4.2). Then we have the following equality:

$$\left(g\left(t - \frac{x}{c}\right) - \frac{k\gamma}{2} g^2\left(t - \frac{x}{c}\right) \right)' - \frac{1}{c} g'\left(t - \frac{x}{x}\right)$$
$$+ \frac{\gamma}{2c} R^* g'\left(t - \frac{x}{c}\right) = 0$$

whence, by integrating over t on $(-\infty, t)$, we obtain

$$\left(1 - \frac{1}{c}\right) g\left(t - \frac{x}{c}\right) - \frac{k\gamma}{2} g^2\left(t - \frac{x}{x}\right)$$
$$+ \frac{\gamma}{2c} \int_{-\infty}^{t} R(t - \tau) g(\tau - \frac{x}{c}) d\tau = \text{const} \,.$$

By virtue of (3.4.5,6), the previous equality can obviously be rewritten as

$$\frac{k\gamma}{2} g(z) \left(g(0) - g(z)\right) + \frac{\gamma}{2} \left(1 - \frac{k\gamma}{2} g(0)\right) \int_{0}^{z} R(z - \xi) g(\xi) d\xi = \text{const}$$

where $z = t - x/c$; it is clear that we can consider $z \geq 0$.

Setting $z = 0$ in the last equality, we obtain const $= 0$. By redenoting z by t, we finally arrive at the equation

3. Nonlinear Waves in Media with Memory

$$g(t)(g(t) - g(0)) - A_0 \int_0^t R(t-\tau)g(\tau)d\tau = 0 ; \quad t \geq 0, \tag{3.4.8}$$

where

$$A_0 \equiv \frac{1 - k\gamma g(0)/2}{k}. \tag{3.4.9}$$

It is supposed that the small parameter γ is sufficiently small for $A_0 > 0$.

Note the nonstandard, **self-coordinated** character of (3.4.8): the quantity $g(0)$ which is the "initial condition" for the function $g(t)$ directly enters (3.4.8).

3.4.3 Estimate of the Solution of the Integral Equation

Lemma 3.4.3. Let for some value $g(0) > 0$ the equation (3.4.8) have a bounded monotonically increasing solution $g(t)$, $t \geq 0$. Then

$$g(t) \to g(\infty) = g(0) + A_0 J \quad \text{as} \quad t \to \infty \tag{3.4.10}$$

where $J \equiv \int_0^\infty R(\tau)d\tau$, the inequalities

$$\frac{g(0)}{2} + \sqrt{\frac{g^2(0)}{4} + A_0 g(0) \int_0^t R(\tau)d\tau} \leq g(t)$$

$$\leq g(0) + A_0 \int_0^t R(\tau)d\tau \tag{3.4.11}$$

hold and the function $f(t)$ is continuous for $t \geq 0$.

Proof. Passing to the limit in (3.4.8) as $t \to \infty$, we easily obtain

$$g(\infty)(g(\infty) - g(0)) = A_0 g(\infty) J$$

whence (3.4.10) follows immediately (since $g(\infty) \neq 0$ by virtue of our assumption that the function $g(t)$ is increasing).

Furthermore, from (3.4.8) one easily obtains

$$g(t)(g(t) - g(0)) \leq A_0 g(t) \int_0^t R(\tau)d\tau$$

whence the right-hand inequality in (3.4.11) follows.

On the other hand, from (3.4.8) it follows that

$$g(t)(g(t) - g(0)) \geq A_0 g(0) \int_0^t R(\tau)d\tau$$

which gives the left-hand inequality in (3.4.11).

Let us now prove the continuity of $g(t)$ for $t \geq 0$. By virtue of strict positiveness of $g(t)$, which is evident from the conditions of the lemma, it follows from (3.4.8) that

$$g(t) = \frac{g(0)}{2} + \sqrt{\frac{g^2(0)}{4} + A_0 \int_0^t R(t-\tau)g(\tau)d\tau} \ . \tag{3.4.12}$$

It also follows from the conditions of the lemma that the integral summand under the radical sign on the right-hand side of (3.4.12) is a continuous function for $t \geq 0$. Therefore, it follows from (3.4.12) that $g(t)$ is continuous for $t \geq 0$, which completes the proof.

Remark. If the kernel $R(t)$ is infinitely differentiable for $t > 0$ and, in addition,

$$|R^{(j)}(0)| < \infty; \quad j = 0, 1, \ldots,$$

then the function $g(t)$ also proves to be infinitely differentiable for $t > 0$. The validity of the fact is obvious from (3.4.8).

3.4.4 Existence of Stationary Profile Shock-Waves

In this section we use a variant of the method of successive approximations which is somewhat different from that of Sects. 3.3.4,5 and directly prove the existence of the solution $g(t)$ (mentioned in Lemma 3.4.3) for the kernel $R(t)$ of a general form. Special cases of the power and exponential kernels will be discussed as examples below.

Theorem 3.4.4 [3.13]. For each value $g(0) > 0$ the equation (3.4.8) has a bounded monotone increasing solution $g(t)$, $t \geq 0$. (Obviously, all the assertions of Lemma 3.4.3 are valid for this solution.)

Proof. Let us construct the solution of (3.4.8) by the method of successive approximations by setting

$$g_n(t)(g_n(t) - g(0)) = A_0 \int_0^t R(t-\tau)g_{n-1}(\tau)d\tau, \quad t \geq 0 \tag{3.4.13}$$

where $g_0(t)$ is an arbitrary continuous positive increasing function, such that $g_0(0) = g(0) > 0$, $g_0(\infty) = g(0) + A_0 J$. At each step here we use for the solution of (3.4.13) the greater root, namely,

$$g_n(t) = \frac{g(0)}{2} + \sqrt{\frac{g^2(0)}{4} + A_0 \int_0^t R(t-\tau)g_{n-1}(\tau)d\tau} \ . \tag{3.4.13'}$$

It is evident from (3.4.13') that $g_n(0) = g(0)$, $g_n(t) \geq g(0) > 0$; $n = 1, 2, \ldots$. In addition, by using (3.4.13') it is easy to prove by induction that all $g_n(t)$ are increasing functions.

Let us now show that the sequence $g_n(t)$ converges for $0 \leq t \leq \delta$ where δ is sufficiently small. Substracting from (3.4.13) the analogous equation where n is replaced by $n - 1$, we obtain

$$(g_n(t) - g_{n-1}(t))(g_n(t) + g_{n-1}(t)) - g(0)(g_n(t) - g_{n-1}(t))$$
$$= A_0 \int_0^t R(t-\tau)(g_{n-1}(\tau) - g_{n-2}(\tau))\, d\tau$$

whence,

$$|g_n(t) - g_{n-1}(t)|(g_n(t) + g_{n-1}(t) - g(0))$$
$$= A_0 \int_0^t R(t-\tau)|g_{n-1}(\tau) - g_{n-2}(\tau)|\, d\tau \, .$$

Therefore, by virtue of monotonic increase and nonnegativeness of functions $g_i(t)$, we have

$$|g_n(t) - g_{n-1}(t)| \leq \frac{A_0}{g(0)} \int_0^t R(t-\tau)|g_{n-1}(\tau) - g_{n-2}(\tau)|\, d\tau$$

whence, taking into account that $0 \leq t \leq \delta$, we obtain

$$|g_n(t) - g_{n-1}(t)| \leq \frac{A_0 \int_0^\delta R(\tau)d\tau}{g(0)} \max_{0\leq\tau\leq\delta} |g_{n-1}(\tau) - g_{n-2}(\tau)| \, . \qquad (3.4.14)$$

A stronger inequality directly follows from the previous one:

$$\max_{0\leq\tau\leq\delta} |g_n(\tau) - g_{n-1}(\tau)|$$
$$\leq \frac{A_0 \int_0^\delta R(\tau)d\tau}{g(0)} \max_{0\leq\tau\leq\delta} |g_{n-1}(\tau) - g_{n-2}(\tau)| \, . \qquad (3.4.15)$$

The inequality (3.4.15) implies the uniform convergence of successive approximations on the segment $[0, \delta]$ under the condition

$$\frac{A_0 \int_0^\delta R(\tau)d\tau}{g(0)} < 1 \, .$$

The previous condition, however, will be obviously satisfied if $\delta > 0$ is taken sufficiently small. Thus, on some small segment $[0, \delta]$ there exists a limiting function

$$\tilde{g}(t) = \lim_{n\to\infty} g_n(t) \, .$$

It is clear that $\tilde{g}(0) = g(0)$ [since all $g_n(0) = g(0)$] and the constructed function $\tilde{g}(t)$ satisfies the equation (3.4.8) for $0 \leq t \leq \delta$.

Furthermore, it is obvious that the function $\tilde{g}(t)$ is positive and monotonically increasing on $[0, \delta]$.

Now, to prove that the constructed solution $\tilde{g}(t)$ can be extended from the small segment $[0, \delta]$ to the entire semi-axis $t \geq 0$, we make use of a refined variant of the successive approximations method, as was used when proving Theorem 3.3.4.

Namely, we set

$$g_0(t) = \tilde{g}(t) \quad \text{for} \quad 0 \le t \le \delta,$$
$$g_0(t) = g(0) + A_0 J \quad \text{for} \quad t \ge \delta \tag{3.4.16}$$

and define the successive approximations $g_n(t)$ by the same formula (3.4.13). Now, in the manner of Theorem 3.3.4 we easily obtain that all $g_n(t)$ are monotonically increasing functions, and

$$g_n(t) = g_0(t) = \tilde{g}(t) \quad \text{for} \quad 0 \le t < \delta,$$
$$g_n(t) \to g(0) + A_0 J \quad \text{as} \quad t \to \infty$$

and the inequalities

$$0 < g_n(t) \le g_{n-1}(t) \le \ldots \le g_0(t)$$

hold on the entire semi-axis $t \ge 0$. Just like in the proof of Theorem 3.3.4, this allows us to conclude that as $n \to \infty$, the sequence $g_n(t)$ converges for $t \ge 0$ to some limiting function which will now be denoted by $g(t)$. Obviously, $g(t)$ is the desired solution of (3.4.8). The theorem is proved.

Let us now turn to the study of special cases.

3.4.5 The Power Kernel

Let

$$R(t) = Ct^{\alpha-1} \quad \text{for} \quad 0 < t < t_0 \tag{3.4.17}$$

where $C > 0$, $0 < \alpha \le 1$. Then for $0 \le t < t_0$ (3.4.8) will take the form

$$g(t)(g(t) - g(0)) = C_1 \int_0^t (t-\tau)^{\alpha-1} g(\tau) d\tau \tag{3.4.18}$$

where

$$C_1 = \frac{C}{k}\left(1 - \frac{k\gamma}{2} g(0)\right).$$

For small $t > 0$ we shall seek the solution of (3.4.18) in the form

$$g(t) = g(0) + g_1 t^\alpha + g_2 t^{2\alpha} + \ldots . \tag{3.4.19}$$

We have, by inserting (3.4.19) into (3.4.18),

$$\{g(0) + g_1 t^\alpha + g_2 t^{2\alpha} + \ldots\}\{g_1 t^\alpha + g_2 t^{2\alpha} + \ldots\}$$
$$= C_1 \int_0^t (t-\tau)^{\alpha-1} g(0) d\tau + C_1 \int_0^t (t-\tau)^{\alpha-1} g_1 \tau^\alpha d\tau + \ldots . \tag{3.4.20}$$

Now, because

$$\int_0^t (t-\tau)^{\alpha-1}\tau^{n\alpha}d\tau = \frac{\Gamma(\alpha)\Gamma(n\alpha+1)}{\Gamma((n+1)\alpha+1)}\, t^{(n+1)\alpha}, \quad n=1,2,\ldots$$

(where Γ is the Euler's Gamma function), we arrive at the following formula by equating the coefficients to equal powers of t in (3.4.20):

$$g(0)g_1 t^\alpha = \frac{C_1 \Gamma(\alpha)\Gamma(1)}{\Gamma(\alpha+1)}\, g(0)t^\alpha,$$

$$\left(g_1^2 + g(0)g_2\right)t^{2\alpha} = \frac{C_1 \Gamma(\alpha)\Gamma(\alpha+1)}{\Gamma(2\alpha+1)}\, g_1 t^{2\alpha}, \tag{3.4.21}$$

$\ldots$

One can see from (3.4.21) that the value of the jump $g(0) > 0$ can be specified arbitrarily (which is in accordance with the result of Theorem 3.4.1), and the coefficients $g_1, g_2, \ldots$ can be found recurrently. In particular,

$$g_1 = \frac{C_1}{\alpha} \equiv \frac{C}{k\alpha}\left(1 - \frac{k\gamma}{2}g(0)\right).$$

Thus, for small $t \geq 0$

$$g(t) = g(0) + \frac{C_1}{\alpha} t^\alpha + \ldots. \tag{3.4.22}$$

3.4.6 The Exponential Kernel

Of interest is also the case where the exponential kernel is defined on the entire semi-axis, i.e.,

$$R(t) = Ce^{-\alpha t}, \quad t \geq 0 \tag{3.4.23}$$

where $C > 0$; $\alpha > 0$. Then (3.4.8) takes the form

$$g(t)\left(g(t) - g(0)\right) = C_1 \int_0^t e^{-\alpha(t-\tau)} g(\tau) d\tau \tag{3.4.24}$$

where

$$C_1 = \frac{C}{k}\left(1 - \frac{k\gamma}{2}g(0)\right).$$

Let us now differentiate (3.4.24) with respect to t. We have

$$g'(t)\left(2g(t) - g(0)\right) = -C_1 \alpha \int_0^t e^{-\alpha(t-\tau)}g(\tau)d\tau + C_1 g(t). \tag{3.4.25}$$

Eliminating the integral term form (3.4.24,25), we arrive at the ordinary differential equation

$$\alpha g(g - g(0)) + g'(2g - g(0)) = C_1 g$$

whence

$$\left(g - \frac{g(0)}{2}\right)\frac{dg}{dt} = \frac{\alpha g}{2}\left(\frac{C_1}{\alpha} + g(0) - g\right). \tag{3.4.26}$$

From (3.4.9,10) it is easy to see that

$$\frac{C_1}{\alpha} + g(0) = g(\infty) > \frac{g(0)}{2}; \tag{3.4.27}$$

thus (3.4.26) finally assumes the form

$$\frac{dg}{dt} = \frac{\alpha}{2}\frac{g(g(\infty) - g)}{g - g(0)/2}. \tag{3.4.28}$$

The solution of (3.4.28), which assumes the value $g(0)$ for $t = 0$ is, evidently, the following:

$$\frac{2}{\alpha}\int_{g(0)}^{g} \frac{g - g(0)/2}{g(g(\infty) - g)}\, dg = t$$

whence by integrating we obtain

$$(g(\infty) - g)^{-1 + g(0)/2g(\infty)} g^{-g(0)/2g(\infty)} = \text{const}\exp(\alpha t/2) \tag{3.4.29}$$

where

$$\text{const} = (g(\infty) - g(0))^{-1 + g(0)/2g(\infty)} (g(0))^{-g(0)/2g(\infty)}.$$

In this connection see *Nigul* and *Stulov* [3.15].

Problem. Reduce (3.4.8) with the oscillatory kernel (3.3.39) to an ordinary differential equation of the second order. In what cases can this equation be analytically integrated?

3.4.7 Waves Propagating in a Prestressed Medium

Now we shall briefly touch upon the case where the medium in front of the wave front is disturbed. Our calculations here will be completely analogous to that of Sect. 3.3.9. As to kernels of general type, we again shall be able to obtain only nonstrict qualitative results. In the case of exponential kernels these results can be rigorously proved; we leave this to the reader.

Thus, we again seek the generalized solution of (3.4.2) in the form $\sigma = g(t - x/c)$, but now with

$$c = \left(1 - \frac{k\gamma}{2}\frac{[g^2]}{[g]}\right)^{-1}. \tag{3.4.30}$$

As in Sect. 3.4.2, we arrive at the equation

$$\left(1 - \frac{1}{c}\right) g\left(t - \frac{x}{c}\right) - \frac{k\gamma}{2} g^2\left(t - \frac{x}{c}\right)$$

$$+ \frac{\gamma}{2c} \int_{-\infty}^{t} R(t - \tau) g(\tau - \frac{x}{c}) d\tau = \tilde{C}_0 , \quad \tilde{C}_0 = \text{const}$$

whence, by redenoting $z = t - x/c$ by t and taking into account (3.4.30), we finally obtain

$$g^2(t) - \frac{[g^2]}{[g]} g(t) - \frac{1 - \frac{k\gamma}{2} \frac{[g^2]}{[g]}}{k} \int_{-\infty}^{t} R(t - \tau) g(\tau) d\tau = C_0 \qquad (3.4.31)$$

where

$$C_0 = -\frac{2}{k\gamma} \tilde{C}_0 .$$

As in (3.4.8), the nonlinear integral equation (3.4.31) is evidently self-coordinated.

To be exact, we again consider that the shock-wave front passes through the origin of coordinates at the moment $t = 0$ and the function g is right continuous (i.e., $g(+0) = g(0)$). Then it is evident that

$$\frac{[g^2]}{[g]} = g(0) + g(-0) .$$

Since k is positive, the stability condition for the shock-wave is given by the following inequality [see (3.2.16')]:

$$[g] = g(0) - g(-0) > 0 . \qquad (3.4.32)$$

Suppose (3.4.31) has a solution tending to finite limits as $t \to \pm\infty$. We denote these limits by $g(\infty)$ and $g(-\infty)$, respectively. The arguments as in Sect. 3.3.9 make it clear that the following inequality must hold:

$$g(\infty) > g(-\infty) . \qquad (3.4.33)$$

In (3.4.31) we now allow t to approach $\pm\infty$. Taking into account the fact that the integral

$$J = \int_0^{\infty} R(\tau) d\tau$$

is finite, we easily obtain the following relations for the limiting values $g(\infty)$ and $g(-\infty)$:

$$g(\infty) - \frac{[g^2]}{[g]} g(\infty) - \frac{1 - \frac{k\gamma}{2} \frac{[g^2]}{[g]}}{k} J g(\infty) = C_0 , \qquad (3.4.34)$$

$$g(-\infty) - \frac{[g^2]}{[g]} g(-\infty) - \frac{1 - \frac{k\gamma}{2}\frac{[g^2]}{[g]}}{k} Jg(-\infty) = C_0 \,. \tag{3.4.34'}$$

Thus, $g(\infty)$ and $g(-\infty)$ appear to be two distinct real roots of the same quadratic equation, which implies that the discriminant of this equation must be positive, i.e.,

$$\frac{1}{4}\left(\frac{J}{k} + \left(1 - \frac{\gamma J}{2}\right)\frac{[g^2]}{[g]}\right) + C_0 > 0 \,.$$

Assuming the last inequality to be satisfied, let us eliminate the constant C_0 from (3.4.34) and (3.4.34'). By a calculation similar to that of Sect. 3.3.9 we easily obtain

$$g(\infty) + g(-\infty) = \frac{J}{k} + \left(1 - \frac{\gamma J}{2}\right)\frac{[g^2]}{[g]} \,.$$

From this relation and (3.4.30) the formula for the velocity of the shock-wave follows

$$c = \frac{1 - \frac{\gamma J}{2}}{1 - \frac{k\gamma}{2}(g(\infty) + g(-\infty))} \,. \tag{3.4.35}$$

This formula obviously follows the structure of (3.3.47).

Problem. Let the kernel of memory be the exponential function (3.4.23).
a) Prove that inequalities (3.4.33) and $g(\infty) > (1-\gamma J)g(-\infty)+J/k$, if satisfied, are the condition of existence of the shock in the wave of steady profile.
b) Calculate $g(0)$ and $g(-0)$ by means of $g(\infty)$ and $g(-\infty)$.

Remark. One can become familiar with a different approach to the problems of stationary profile wave propagation in the works by *Nigul* et al. [3.14,15]. Interesting results are also found in [3.8].

3.5 Waves Tending to a Stationary Profile

3.5.1 Intuitive Approach

In Sects. 3.3,4 we dealt with continuous and discontinuous generalized solutions of the equation

$$\frac{\partial}{\partial t}\left(\sigma - \frac{k\gamma}{2}\sigma^2\right) + \left(1 - \frac{\gamma}{2}R^*\right)\frac{\partial \sigma}{\partial x} = 0 \tag{3.5.1}$$

which describe waves of stationary profile travelling to the right in the undisturbed medium. Let us recall that here $k > 0$, $0 < \gamma \ll 1$ and

$$R^*u \equiv \int_{-\infty}^{t} R(t-\tau)u(\tau)d\tau$$

where $R(t)$ is a nonnegative decreasing function such that

$$J \equiv \int_{0}^{\infty} R(t)dt < \infty.$$

As we know, continuous stationary profile waves which travel in the undisturbed medium have the form

$$\sigma = f(t - x + \text{const})$$

where $f(z) = 0$ for $z \leq 0$, and discontinuous ones have the form

$$\sigma = g(t - x/c + \text{const})$$

where $g(z) = 0$ for $z < 0$; $g(0) > 0$,

$$c = \left(1 - \frac{k\gamma}{2} g(0)\right)^{-1}.$$

Let us compare the formula for $g(\infty)$ (3.4.9,10)

$$g(\infty) = \frac{J}{k} + g(0)\left(1 - \frac{\gamma J}{k}\right) \tag{3.5.2}$$

with the formula for $f(\infty)$ (3.3.6):

$$f(\infty) = \frac{J}{k}. \tag{3.5.3}$$

It is clear that under our assumptions

$$g(\infty) > f(\infty) > 0. \tag{3.5.4}$$

Moreover, the value $g(\infty)$ increases with $g(0)$.

From (3.5.2–4) one can make some qualitative conclusions concerning the large time behaviour of the (generalized) nonstationary solution of (3.5.1) under the following initial and boundary conditions:

$$\begin{aligned}&\sigma = 0 \quad \text{for} \quad x > 0, \quad t \leq 0; \\ &\sigma(t,0) = \sigma_0 \Theta(t).\end{aligned} \tag{3.5.5}$$

Here

$$\sigma_0 > 0, \quad \begin{aligned}\Theta(t) &= 1 \quad \text{for} \quad t \geq 0, \\ &= 0 \quad \text{for} \quad t < 0.\end{aligned}$$

a) If $\sigma_0 > J/k$, then as $t \to \infty$ the solution of the problem (3.5.1,5) will tend to a stationary profile shock-wave which travels in the undisturbed medium, i.e.,

$$\sigma = g(t - x/c + \text{const}) .$$

Here $g(z)$ is a solution of (3.4.8) for $z \geq 0$; $g(z) \equiv 0$ for $z < 0$. The wave front velocity c and the value of the jump on the front $g(0)$ can be found from the equality

$$c = \left(1 - \frac{k\gamma}{2} g(0)\right)^{-1} \tag{3.5.6}$$

(3.4.6) together with an intuitively clear relation

$$\sigma_0 = g(\infty) .$$

By virtue of (3.5.2) that relates $g(\infty)$ with $g(0)$, the previous equality takes the form

$$\sigma_0 = \frac{J}{k} + g(0)\left(1 - \frac{\gamma J}{2}\right) . \tag{3.5.7}$$

From (3.5.6,7) we obtain

$$g(0) = \frac{\sigma_0 - \frac{J}{k}}{1 - \frac{\gamma J}{2}} ,$$

$$c = \frac{1 - \frac{\gamma J}{2}}{1 - \frac{k\gamma\sigma_0}{2}} > 1 . \tag{3.5.8}$$

One can also obtain the latter formula from (3.4.35) by putting $g(\infty) = \sigma_0$, $g(-\infty) = 0$. The presence of a nonfading jump on the front $[\sigma] = g(0)$ indicates the fact that the nonlinear effect of overtaking turns out to be stronger than the relaxation. The corresponding wave profile for t large enough is shown in Fig. 3.2a.

b) If $\sigma_0 = J/k$, then as $t \to \infty$ the solution of (3.5.1,5) tends to a continuous stationary profile which travels in the undisturbed medium

$$\sigma = f(t - x + \text{const}) .$$

Here $f(z)$ is a nonzero solution of the integral equation (3.3.5) for $z > 0$; $f(z) \equiv 0$ for $z < 0$. One can say that in this case the nonlinear and relaxational effects get balanced (Fig. 3.2b).

c) If $0 < \sigma_0 < J/k$, then none of the integral equations (3.3.5) or (3.4.8) has a solution which should tend to σ_0 as $t \to \infty$. Therefore, the solution of (3.5.1,5) cannot tend to any of the solutions of these equations. However, it is intuitively clear that the wave profile must tend to some stationary state. Thus, we have to conclude that the solution of (3.5.1,5) tends to a continuous stationary profile wave travelling in the prestressed medium

$$\sigma = f(t - x/c + \text{const}) .$$

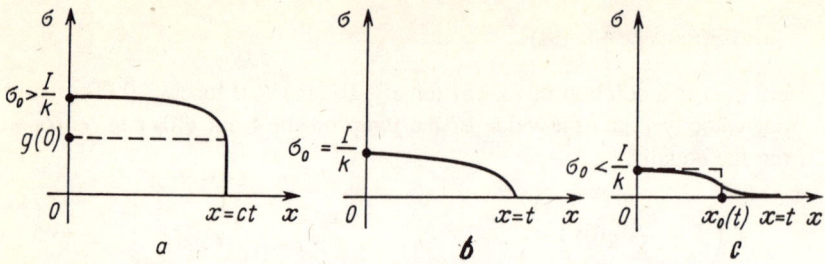

Fig. 3.2. For different cases the function $\sigma(x)$ is plotted, see **a** to **c** from *left* to *right*

Here f is a solution of the integral equation (3.3.43), and the value of the wave velocity c can be determined from (3.3.47') where $f(\infty)$ must be replaced by σ_0, i.e.,

$$c = \frac{1 - \frac{\gamma J}{2}}{1 - \frac{k\gamma\sigma_0}{2}} < 1 . \tag{3.5.9}$$

The corresponding wave profile for large t is shown in Fig. 3.2c. It is clear that the relaxation is a dominant factor in the wave front formation.

3.5.2 Rok's Method

The typical velocity of a wave with a profile, tending to a steady state, can also be estimated in the following way as suggested by *Rok*.

For illustration we take the situation c) considered in Sect. 3.5.1, i.e., $0 < \sigma_0 < J/k$. Integrating (3.5.1) over x for t sufficiently large, we obtain

$$\frac{d}{dt}\int_0^\infty \left(\sigma - \frac{k\gamma}{2}\sigma^2\right)dx = \sigma_0 - \frac{\gamma}{2}\int_0^t R(t-\tau)\sigma_0 d\tau .$$

Let us now approximately replace the genuine wave profile by the rectangular profile of the size $\sigma_0 \times x_0(t)$ (Fig. 3.2c). Then the previous relation evidently yields

$$\left(1 - \frac{k\gamma\sigma_0}{2}\right)\frac{dx_0(t)}{dt} \sim 1 - \frac{\gamma}{2}\int_0^t R(t)dt$$

whence for large t it follows that

$$\frac{dx_0(t)}{dt} \sim \frac{1 - \frac{\gamma}{2}J}{1 - \frac{k\gamma\sigma_0}{2}} .$$

Thus, the typical wave velocity for large t coincides, according to expectations, with (3.5.9), i.e., with the velocity of a wave of stationary profile.

3.6 Nonstationary Waves Analog of the Landau-Whitham Formula

3.6.1 Formulation of the Problem

In Sects. 3.3–5 we dealt with the equation

$$\frac{\partial}{\partial t}\left(\sigma - \frac{k\gamma}{2}\sigma^2\right) + \left(1 - \frac{\gamma}{2}R^*\right)\frac{\partial \sigma}{\partial x} = 0. \tag{3.6.1}$$

As was already pointed out (Sect. 3.2), this equation can be rewritten with accuracy to $O(\gamma^2)$ in the form

$$\frac{\partial}{\partial t}\left(\sigma - \frac{k\gamma}{2}\sigma^2\right) + \frac{\partial \sigma}{\partial x} + \frac{\gamma}{2}R^*\frac{\partial \sigma}{\partial t} = 0 \tag{3.6.2}$$

or, which is the same [in the domain of smoothness of $\sigma(t,x)$],

$$(1 - k\gamma\sigma)\frac{\partial \sigma}{\partial t} + \frac{\partial \sigma}{\partial x} + \frac{\gamma}{2}R^*\frac{\partial \sigma}{\partial t} = 0. \tag{3.6.2'}$$

In Sect. 3.7 we shall see that there exists a simple modification of the constitutive equation under which the asymptotic equation (3.6.1) proves to be exact. For (3.6.2) we do not know such a modification of the constitutive equation. However, (3.6.2) is more convenient when studying unsteady waves and is therefore used here.

Let us set the follwing problem for (3.6.2):

$$\begin{aligned}\sigma &= 0 \quad \text{for} \quad x > 0, \quad t \leq 0, \\ \sigma(t,0) &= \sigma_0(t)\end{aligned} \tag{3.6.3}$$

where $\sigma_0(t)$ is a smooth function identically equal to zero for $t \leq 0$.

3.6.2 Linear Case

If $k = 0$ (i.e., the constitutive equation is linear), then (3.6.2) takes the form

$$\left(1 + \frac{\gamma}{2}R^*\right)\frac{\partial \sigma}{\partial t} + \frac{\partial \sigma}{\partial x} = 0. \tag{3.6.4}$$

By applying the Laplace transform $L_{t\to p}$ to (3.6.4), we obtain for $x > 0$ that

$$\left(1 + \frac{\gamma}{2}\bar{R}(p)\right)p\bar{\sigma}(p,x) + \frac{d}{dx}\bar{\sigma}(p,x) = 0$$

whence, it obviously follows

$$\bar{\sigma}(p,x) = C(p)\exp\left\{-p\left(1 + \frac{\gamma}{2}\bar{R}(p)\right)x\right\}, \quad x \geq 0.$$

From (3.6.3) it is clear that here

$$C(p) = \bar{\sigma}(p,0) = \bar{\sigma}_0(p).$$

Thus we can finally write

$$\sigma(t,x) = L^{-1}_{p \to t} \bar{\sigma}_0(p) \exp\left(-px - \frac{\gamma}{2} p\bar{R}(p)x\right), \quad x \geq 0 \tag{3.6.5}$$

where $L^{-1}_{p \to t}$ is the inverse Laplace transform. It easily follows from the properties of the Laplace transform that the previous equality can be rewritten as

$$\sigma(t,x) = L^{-1}_{p \to t-x} \bar{\sigma}_0(p) \exp\left(-\frac{\gamma}{2} p\bar{R}(p)x\right), \quad x \geq 0. \tag{3.6.5'}$$

3.6.3 Case of Small Quadratic Nonlinearity

Now let $k \neq 0$. At first we suppose that there are no strong shocks; then by analogy with the Landau–Whitham approach [3.16,17] we define the function

$$\sigma(t,x) = L^{-1}_{p \to \tau} \bar{\sigma}_0(p) \exp\left(-\frac{\gamma}{2} p\bar{R}(p)x\right), \quad x \geq 0 \tag{3.6.6}$$

where "the phase" $\tau = \tau(t,x)$ is determined by the condition of its being constant along the characteristics

$$\frac{dt}{dx} = 1 - k\gamma\sigma \quad \text{for} \quad \tau = \text{const}, \tag{3.6.7}$$

$$\tau\big|_{x=0} = t. \tag{3.6.8}$$

It is clear that the right-hand side of (3.6.6) is a known function of the variable τ and x; we denote it by $G(\tau,x)$. Substituting this function (instead of σ) into (3.6.7), we obtain

$$\frac{dt}{dx} = 1 - k\gamma G(\tau,x)$$

whence, because of (3.6.8), we have

$$t = \int_0^x (1 - k\gamma G(\tau,y))dy + \tau. \tag{3.6.9}$$

It is (3.6.9) that determines τ as a function of t, x.

It is also obvious that in the linear case the approximation (3.6.6–8) turns into (3.6.5').

3.6.4 Estimate of Quality of the Approximate Solution

Let us now observe the way the approximate solution (3.6.6–8) works in the nonlinear situation. If $R(t) = 0$, then (3.6.6–8) obviously yield

$$\sigma(t, x) = \sigma_0(\tau),$$
$$t = (1 - k\gamma\sigma_0(\tau))x + \tau. \qquad (3.6.10)$$

A simple checking shows that (3.6.10) proves to be an exact solution of the equation

$$(1 - k\gamma\sigma)\frac{\partial\sigma}{\partial t} + \frac{\partial\sigma}{\partial x} = 0$$

and satisfies the initial and boundary conditions (3.6.3).

Consider now a more general case where

$$R(t) = R(0) = \text{const}. \qquad (3.6.11)$$

In this case

$$\bar{R}(p) = \frac{R(0)}{p}$$

and from (3.6.6) it follows that

$$\sigma(t, x) = \exp\left(-\frac{\gamma}{2} R(0)x\right) \sigma_0(\tau). \qquad (3.6.12)$$

Substituting (3.6.12) into (3.6.7) and taking into account (3.6.8), we obtain

$$t = \int_0^x \left(1 - k\gamma \exp\left(-\frac{\gamma}{2} R(0)y\right) \sigma_0(\tau)\right) dy + \tau \qquad (3.6.13)$$

whence τ can be determined as a function of t, x.

Let us show that again (3.6.12,13) turns out to be an exact solution of the corresponding single-wave equation! In fact, in case (3.6.11) the equation (3.6.2') takes the form

$$(1 - k\gamma\sigma)\frac{\partial\sigma}{\partial t} + \frac{\partial\sigma}{\partial x} + \frac{\gamma}{2} R(0)\sigma = 0.$$

The solution of the last equation is determined by the equations of characteristics

$$\frac{dt}{dx} = 1 - k\gamma\sigma, \quad \text{and} \quad \frac{d\sigma}{dx} = -\frac{\gamma}{2} R(0)\sigma.$$

Integrating these equations and taking into account (3.6.3), we arrive exactly at the relation (3.6.12,13).

One might assume that (3.6.6–8) give, in general, the exact solution to the posed single-wave problem (3.6.2,3). However, that is not the case. One can check that (3.6.6–8) is only a good asymptotic approximation with accuracy of order $O(\gamma^2)$.

Problem. Making use of (3.6.6–8), estimate the wave breaking coordinates.

Finally, if (3.6.6) becomes multi-valued during the process of motion, this implies the formation of a shock wave. One may approximately construct the shock-wave front, for example, using the method of Sect. 1.5.2.

Remark. The approach given above for solving (3.6.2) is, evidently, applicable to the more general equation

$$\frac{\partial f(\sigma)}{\partial t} + \frac{\partial \sigma}{\partial x} + \frac{\gamma}{2} R^* \frac{\partial \sigma}{\partial t} = 0 \qquad (3.6.14)$$

where $f(0) = 0$, $f'(\sigma) > 0$.

3.6.5 Single-Wave Equation for Deformation

We now derive the single-wave equation for deformation. First of all, recall that the constitutive equation for the medium under consideration has the form

$$\varepsilon = \frac{1}{1 - \gamma R^*}(A\sigma + \gamma B\sigma^2). \qquad (3.6.15)$$

Solving this equation for σ, we easily obtain [with accuracy to terms of order $O(\gamma^2)$]

$$\sigma = \frac{1 - \gamma R^*}{A}\varepsilon - \frac{\gamma B}{A^3}\varepsilon^2,$$

i.e.,

$$\sigma = \frac{1 - \gamma R^*}{A}\varepsilon + \frac{k\gamma}{A^2}\varepsilon^2, \quad k \equiv -\frac{B}{A}. \qquad (3.6.16)$$

We insert the obtained expression for σ into (3.6.2'). By ignoring again the terms of order $O(\gamma^2)$, we easily obtain the desired single-wave equation for deformation in the form analogous to (3.6.2'):

$$(1 - k_1\gamma\varepsilon)\frac{\partial \varepsilon}{\partial t} + \frac{\partial \varepsilon}{\partial x} + \frac{\gamma}{2} R^* \frac{\partial \varepsilon}{\partial t} = 0, \quad k_1 \equiv -\frac{B}{A^2}. \qquad (3.6.17)$$

Remark. From (3.6.17) it obviously follows that $\partial \varepsilon/\partial t = -\partial \varepsilon/\partial x + O(\gamma)$. Hence one can easily reduce [with accuracy to terms of order $O(\gamma^2)$] (3.6.17) to

$$\left(1 + \frac{\gamma}{2} R^*\right)\frac{\partial \varepsilon}{\partial t} + \frac{\partial}{\partial x}\left(\varepsilon + \frac{k_1\gamma}{2}\varepsilon^2\right) = 0. \qquad (3.6.18)$$

This form of the equation for deformation will be useful below in deriving the single-wave equation for displacement.

3.6 Nonstationary Waves Analog of the Landau-Whitham Formula

Let us now set the following boundary value problem for deformation:

$$\varepsilon = 0 \quad \text{for} \quad x > 0, \quad t \leq 0,$$
$$\varepsilon(t, 0) = \varepsilon_0(t) \tag{3.6.19}$$

where $\varepsilon_0(t)$ is a smooth function identically equal to zero for $t \leq 0$. Then, the asymptotic solution of (3.6.17,19) is obviously presented by the formula analogous to (3.6.6):

$$\varepsilon(t, x) = L^{-1}_{p \to \tau} \bar{\varepsilon}_0(p) \exp\left(-\frac{\gamma}{2} p \bar{R}(p) x\right) \tag{3.6.20}$$

where $\tau = \tau(t, x)$ is determined by its contancy along the characteristics:

$$\frac{dt}{dx} = 1 - k_1 \gamma \varepsilon \quad \text{for} \quad \tau = \text{const}; \quad \tau \bigg|_{x=0} = t.$$

3.6.6 Single-Wave Equation for Displacement

We now derive the single-wave equation for displacement. In this connection we recall that the variable x which is used here is, in fact, the coordinate $\bar{x}$ introduced in Sect. 3.2.5, the bar over it being omitted for simplicity. Thus, we have

$$\varepsilon = \sqrt{A\varrho} \, \frac{\partial u}{\partial x}. \tag{3.6.21}$$

We prefer the form (3.6.18) of the equation for deformation because it contains the derivative $\partial \varepsilon / \partial t$ without a nonlinear factor. Inserting (3.6.21) into (3.6.18), we obtain

$$\left(1 + \frac{\gamma}{2} R^*\right) \frac{\partial}{\partial t} \sqrt{A\varrho} \, \frac{\partial u}{\partial x} + \frac{\partial}{\partial x} \left\{ \sqrt{A\varrho} \, \frac{\partial u}{\partial x} + \frac{k_1 \gamma}{2} \left(\sqrt{A\varrho} \, \frac{\partial u}{\partial x}\right)^2 \right\} = 0$$

whence, by commutativity of the operators of convolution and differentiation, it follows that

$$\sqrt{A\varrho} \frac{\partial}{\partial x} \left\{ \left(1 + \frac{\gamma}{2} R^*\right) \frac{\partial u}{\partial t} + \frac{\partial u}{\partial x} + \frac{k_1 \gamma}{2} \sqrt{A\varrho} \left(\frac{\partial u}{\partial x}\right)^2 \right\} = 0.$$

The last equality obviously yields

$$\left(1 + \frac{\gamma}{2} R^*\right) \frac{\partial u}{\partial t} + \frac{\partial u}{\partial x} + \frac{k_1 \gamma}{2} \sqrt{A\varrho} \left(\frac{\partial u}{\partial x}\right)^2 = f(t) \tag{3.6.22}$$

where $f(t)$ is some indeterminate function.

3.6.7 A Boundary Value Problem Posed in Terms of Displacement

Now, let us set the following problem for displacement:

$$u = 0 \quad \text{for} \quad x > 0, \quad t \leq 0,$$
$$u(t,0) = u_0(t) \tag{3.6.23}$$

where $u_0(t)$ is a smooth function identically equal to zero for $t \leq 0$. It is clear that for $x = \text{const} > 0$ sufficiently large $u(t,x)$ equals zero within an arbitrarily long interval of time. Hence, in the case under consideration we must set $f \equiv 0$ in (3.6.22), and the single-wave equation for displacement finally assumes the form

$$\left(1 + \frac{\gamma}{2} R^*\right) \frac{\partial u}{\partial t} + \frac{\partial u}{\partial x} + \frac{k_1 \gamma}{2} \sqrt{A\varrho} \left(\frac{\partial u}{\partial x}\right)^2 = 0. \tag{3.6.24}$$

To construct the asymptotic solution of (3.6.23,24) formulated in terms of displacement, we reformulate it in terms of deformation (in the manner of Sect. 1.3.7). Letting $x = 0$ in (3.6.24) and using (3.6.21), we obtain a quadratic equation for $\varepsilon(t,0)$:

$$\left(1 + \frac{\gamma}{2} R^*\right) u_0'(t) + \frac{\varepsilon(t,0)}{\sqrt{A\varrho}} + \frac{k_1 \gamma}{2} \sqrt{A\varrho} \left(\frac{\varepsilon(t,0)}{\sqrt{A\varrho}}\right)^2 = 0 \tag{3.6.25}$$

whence, after denoting $\varepsilon(t,0)$ by $\varepsilon_0(t)$, we have

$$\varepsilon_0(t) = \frac{-1 + \sqrt{1 - 2k_1\gamma\sqrt{A\varrho}\left(1 + \frac{\gamma}{2} R^*\right) u_0'(t)}}{k_1\gamma}. \tag{3.6.26}$$

We have chosen here the greater of the two roots of (3.6.25) because as $\gamma \to 0$ the function $\varepsilon_0(t)$ must remain bounded (and pass into the corresponding function of the linear elastic problem). It is clear that (3.6.26) is equal to zero for $t \leq 0$. Then, it is obvious from (3.6.23) that $\varepsilon = 0$ for $x > 0$, $t \leq 0$. It is also obvious that the travelling deformation wave is determined by the single-wave equation (3.6.18) or, which is the same [with accuracy to terms of order of $O(\gamma^2)$], by (3.6.17). To establish this fact, it is enough to differentiate (3.6.24) with respect to x. Therefore, in the problem under consideration the deformation wave is asymptotically determined by (3.6.20), where the function $\varepsilon_0(t)$ is expressed by (3.6.26). Finally, since the deformation wave is found, one can find the displacement in our problem:

$$u(t,x) = \frac{1}{\sqrt{A\varrho}} \int_0^x \varepsilon(t,x) dx + u_0(t). \tag{3.6.27}$$

Remark. Section 3.6 is based on the results of [3.18,19].

3.7 General Nonlinearity. Further Factorization Theorems for Nonlinear Wave Equations with Memory

3.7.1 Preliminary Notes

In the previous sections we were concerned with the constitutive equation (3.2.1) containing small quadratic nonlinearity (and a small hereditary term). Let us now introduce nonlinearity of a general type into the constitutive equation. On the face of it, the following generalization of (3.2.1) seems to be the most convenient (the small parameter γ is left here only as a coefficient to the hereditary summand):

$$\varepsilon(t) - \gamma \int_{-\infty}^{t} R(t-\tau)\varepsilon(\tau)d\tau = a(\sigma) \ .$$

However, we failed to find a way to factorize exactly and even asymptotically, with accuracy $O(\gamma^2)$, the corresponding wave equation for stress,

$$\frac{\partial^2 a(\sigma)}{\partial t^2} - \frac{1}{\varrho}(1 - \gamma R^*)\frac{\partial^2 \sigma}{\partial x^2} = 0, \quad \varrho = \text{const} \ .$$

This failure is caused by the above mentioned noncommutativity of the convolution operator and the operator of multiplication by a function (of (σ)).

To obtain meaningful results concerning interaction of memory and nonlinearity of a general type, we must use a more complicated constitutive equation introduced in Sect. 3.1.2.

3.7.2 The Exact Factorization Theorem

Let σ be a smooth function of t vanishing as $\sigma \to -\infty$ and assume that the constitutive equation (3.1.7) holds, i.e.,

$$\varepsilon = \int_{-\infty}^{t} \sqrt{1 + K^*}\sqrt{a'(\sigma)}\sqrt{1 + K^*}\sqrt{a'(\sigma)}\sigma'_t dt \ . \tag{3.7.1}$$

As was pointed out in Sect. 3.1.2, in the case of $K^* = 0$, (3.7.1) assumes the form

$$\varepsilon = a(\sigma) \ .$$

On the other hand, in the case of $a'(\sigma) = \text{const} = 1/E$, (3.7.1) turns into the well-known constitutive equation of linear hereditary elasticity.

Theorem 3.7.2. [3.13]. Let $\sigma(t, x)$ be a smooth function vanishing as $t \to -\infty$ and suppose that the constitutive equation (3.7.1) holds. Then the corresponding wave equation

$$\frac{\partial^2}{\partial t^2} \int_{-\infty}^{t} \sqrt{1 + K^*}\sqrt{a'(\sigma)}\sqrt{1 + K^*}\sqrt{a'(\sigma)}\sigma'_t dt - \frac{1}{\varrho}\frac{\partial^2 \sigma}{\partial x^2} = 0 \tag{3.7.2}$$

can be exactly factorized as follows:

$$\left\{ \frac{\partial}{\partial t}\sqrt{1+K^*}\sqrt{a'(\sigma)} \mp \frac{1}{\sqrt{\varrho}}\frac{\partial}{\partial x} \right\}$$
$$\times \left\{ \sqrt{1+K^*}\sqrt{a'(\sigma)}\frac{\partial}{\partial t} \pm \frac{1}{\sqrt{\varrho}}\frac{\partial}{\partial x} \right\} \sigma = 0. \tag{3.7.3}$$

Proof. Obviously, it is enough to check that when the expressions in braces in (3.7.3) are directly expanded, one can cancel the summands with mixed derivatives, i.e.,

$$\frac{\partial}{\partial t}\sqrt{1+K^*}\sqrt{a'(\sigma)}\frac{\partial \sigma}{\partial x} - \frac{\partial}{\partial x}\sqrt{1+K^*}\sqrt{a'(\sigma)}\frac{\partial \sigma}{\partial t} = 0. \tag{3.7.4}$$

But, by virtue of commutativity of the convolution operator with differentiation, the left-hand side of (3.7.4) can be rewritten as

$$\sqrt{1+K^*}\left\{ \frac{\partial}{\partial t}\sqrt{a'(\sigma)}\frac{\partial \sigma}{\partial x} - \frac{\partial}{\partial x}\sqrt{a'(\sigma)}\frac{\partial \sigma}{\partial t} \right\}$$
$$\equiv \sqrt{1+K^*}\left\{ \frac{\partial^2}{\partial t \partial x}\int_0^\sigma \sqrt{a'(\sigma)}\,d\sigma - \frac{\partial^2}{\partial x \partial t}\int_0^\sigma \sqrt{a'(\sigma)}\,d\sigma \right\},$$

which gives the result required. The theorem is proved.

Corollary. Let us define the operator Φ^* by the equality

$$1 - \Phi^* \equiv \frac{1}{\sqrt{1+K^*}}. \tag{3.7.5}$$

Then from (3.7.3) it obviously follows that each smooth solution of any of the equations

$$\sqrt{\varrho a'(\sigma)}\frac{\partial \sigma}{\partial t} \pm (1-\Phi^*)\frac{\partial \sigma}{\partial x} = 0 \tag{3.7.6}$$

will be the solution of (3.7.2) as well.

Note that the constitutive equation (3.7.1) is not appropriate for the direct description of strong shocks. Indeed, if σ suffers a strong shock, then the integrand contains a product of the discontinuous function $(a'(\sigma))^{1/2}$ and the function σ'_t having the singularity of the Dirac δ-function type; such a product without any additional admissions is not defined at all. However, it is possible to show that such a regularization of the integral in (3.7.1) exists and that this relation will also make sense for both functions ε and σ suffering strong shocks. Moreover, the above corollary will hold in the complement to the shock-wave front, while across the front the relation

$$[\varepsilon] = [a(\sigma)]$$

will be valid. This relation obviously yields the condition on the shock for the stress wave

$$U = \pm\sqrt{\frac{[\sigma]}{\varrho[a(\sigma)]}}. \tag{3.7.7}$$

However, in the case where σ is a continuous piecewise smooth function, (3.7.1) makes sense and does not need any regularization. This enables us to make the result of the previous theorem somewhat stronger. Let l be the line of the weak discontinuity of σ which is supposed to be (in the complement to l) a solution of (3.7.6), for example, when one is choosing the upper signs. Then, as it easily follows from Theorem 3.7.2, the function σ also satisfies (3.7.2) outside l. It is evident that the line l will be a characteristic of the corresponding single-wave equation.

It is now clear from what point of view the results of Sect. 3.3 can be considered exact. In Sect. 3.3 the continuous solutions of (3.2.9) which evidently belongs to the (3.7.6) type were under consideration. Let us recall that (3.3.9) was obtained as a result of asymptotic factorization (as $\gamma \to 0$). However, one may consider that instead of the constitutive equation (3.2.1),

$$\varepsilon - \gamma R^*\varepsilon = A\sigma + \gamma B\sigma^2$$

the constitutive equation (3.7.1) holds with

$$\sqrt{a'(\sigma)} = \sqrt{A}\left(A + \gamma\frac{B}{A}\sigma\right), \quad \sqrt{1+K^*} = \frac{1}{1-\frac{3}{2}R^*}.$$

Then, by virtue of Theorem 3.7.2, (3.2.9) holds not asymptotically but exactly.

The results of Sect. 3.4 can be considered valid only asymptotically (as $[\sigma] \to 0$) even under this approach. Indeed, we know that for the purely nonlinear constitutive equation $\varepsilon = a(\sigma)$ [which is a special case of the regularized equation (3.7.1)], the presence of strong shocks destroys the supposition of the single-wave character of the process.

Remark. Another approach to the problem considered in this section is given in [3.20].

3.7.3 The Asymptotic Factorization Theorem

Consider now one more nonlinear constitutive equation with memory of a rather special form:

$$\varepsilon = a(\sigma) + \gamma \int_{-\infty}^{t} b(\sigma(\tau))d\tau \tag{3.7.8}$$

where $a(\sigma)$ and $b(\sigma)$ are arbitrary smooth monotonically increasing functions, $a(0) = 0$, γ is a small parameter. The corresponding wave equation for stress has the form

$$\frac{\partial^2}{\partial t^2}\left\{a(\sigma) + \gamma \int_{-\infty}^{t} b(\sigma(\tau, x))d\tau\right\} - \frac{1}{\varrho}\frac{\partial^2 \sigma}{\partial x^2} = 0$$

or, which is the same,

$$\frac{\partial^2 a(\sigma)}{\partial t^2} + \gamma \frac{\partial b(\sigma)}{\partial t} - \frac{1}{\varrho}\frac{\partial^2 \sigma}{\partial x^2} = 0. \tag{3.7.9}$$

Let us try to asymptotically factorize (3.7.9). We shall seek the decomposition of the left-hand side of (3.7.9) in the form

$$\left\{\left(\frac{\partial}{\partial t}\sqrt{a'(\sigma)} + \frac{\gamma}{2}\beta'(\sigma)\right) \mp \frac{1}{\sqrt{\varrho}}\frac{\partial}{\partial x}\right\}$$
$$\times \left\{\left(\sqrt{a'(\sigma)}\frac{\partial \sigma}{\partial t} + \frac{\gamma}{2}\beta(\sigma)\right) \pm \frac{1}{\sqrt{\varrho}}\frac{\partial \sigma}{\partial x}\right\} \tag{3.7.10}$$

where $\beta(\sigma)$ is a function to be determined. By directly expanding the braces in (3.7.10), we obtain

$$\left(\frac{\partial}{\partial t}\sqrt{a'(\sigma)} + \frac{\gamma}{2}\beta'(\sigma)\right)\left(\sqrt{a'(\sigma)}\frac{\partial \sigma}{\partial t} + \frac{\gamma}{2}\beta(\sigma)\right) - \frac{1}{\varrho}\frac{\partial^2 \sigma}{\partial x^2}$$
$$\mp \frac{1}{\sqrt{\varrho}}\left\{\frac{\partial}{\partial x}\left(\sqrt{a'(\sigma)}\frac{\partial \sigma}{\partial t} + \frac{\gamma}{2}\beta(\sigma)\right)\right.$$
$$\left. - \left(\frac{\partial}{\partial t}\sqrt{a'(\sigma)} + \frac{\gamma}{2}\beta'(\sigma)\right)\frac{\partial \sigma}{\partial x}\right\},$$

which identically equals

$$\left(\frac{\partial}{\partial t}\sqrt{a'(\sigma)} + \frac{\gamma}{2}\beta'(\sigma)\right)\left(\sqrt{a'(\sigma)}\frac{\partial \sigma}{\partial t} + \frac{\gamma}{2}\beta(\sigma)\right) - \frac{1}{\varrho}\frac{\partial^2 \sigma}{\partial x^2}. \tag{3.7.10'}$$

Here we have used the identity

$$\frac{\partial}{\partial t}f(\sigma)\frac{\partial \sigma}{\partial x} \equiv \frac{\partial}{\partial x}f(\sigma)\frac{\partial \sigma}{\partial t}$$

which is familiar. By multiplying out the expressions in parentheses in (3.7.10'), we finally obtain the following form of (3.7.10):

$$\frac{\partial^2 a(\sigma)}{\partial t^2} + \left\{\frac{\gamma}{2}\beta'(\sigma)\sqrt{a'(\sigma)}\frac{\partial \sigma}{\partial t} + \frac{\partial}{\partial t}\sqrt{a'(\sigma)}\frac{\gamma}{2}\beta(\sigma)\right\}$$
$$+ \frac{\gamma^2}{4}\beta'(\sigma)\beta(\sigma) - \frac{1}{\varrho}\frac{\partial^2 \sigma}{\partial x^2}. \tag{3.7.11}$$

Let us now determine $\beta(\sigma)$ by requiring that the expression in braces in (3.7.11) be equal to $\gamma \delta b(\sigma)/\partial t$:

$$\frac{\gamma}{2} \beta'(\sigma) \sqrt{a'(\sigma)} \frac{\partial \sigma}{\partial t} + \frac{\partial}{\partial t} \sqrt{a'(\sigma)} \frac{\gamma}{2} \beta(\sigma) = \gamma \frac{\partial b(\sigma)}{\partial t}$$

or, which is the same,

$$\frac{1}{2} \left(2\beta'(\sigma)\sqrt{a'(\sigma)} + \frac{a''(\sigma)}{2\sqrt{a'(\sigma)}} \beta(\sigma) \right) = b'(\sigma)$$

whence,

$$\beta' + \frac{1}{4} \frac{a''(\sigma)}{a'(\sigma)} \beta = \frac{b'(\sigma)}{\sqrt{a'(\sigma)}} . \qquad (3.7.12)$$

The solution of this equation vanishing for $\sigma = 0$ is

$$\beta(\sigma) = (a'(\sigma))^{-1/4} \int_0^\sigma (a'(\sigma))^{-1/4} b'(\sigma) d\sigma . \qquad (3.7.13)$$

It is clear that for $\beta(\sigma)$, defined in (3.7.13), expression (3.7.11) coincides with the left-hand side of (3.7.9) with accuracy to $\gamma^2 \beta'(\sigma)\beta(\sigma)/4$. Thus, we have proved the following result.

Theorem 3.7.3 [3.13]. The equation (3.7.9) can be represented in the form

$$\left\{ \frac{\partial}{\partial t} \sqrt{a'(\sigma)} + \frac{\gamma}{2} \beta'(\sigma) \mp \frac{1}{\sqrt{\varrho}} \frac{\partial}{\partial x} \right\}$$
$$\times \left\{ \sqrt{a'(\sigma)} \frac{\partial \sigma}{\partial t} + \frac{\gamma}{2} \beta(\sigma) \pm \frac{1}{\sqrt{\varrho}} \frac{\partial \sigma}{\partial x} \right\} = O(\gamma^2) \qquad (3.7.14)$$

where the function $\beta(\sigma)$ is defined by (3.7.13); the quantity $O(\gamma^2)$ is uniformly small for bounded σ and vanishes for $\sigma = 0$.

Remark. The single-wave equations that follow from (3.7.13,14) [where one must neglect the discrepancy $O(\gamma^2)$] can be easily solved by the method of characteristics. These single-wave equations correspond to pulse propagation in the undisturbed medium. If we considered the case where the stress pulse propagates in a homogeneously prestressed medium, then instead of (3.7.13) we should, evidently, set

$$\beta(\sigma) = (a'(\sigma))^{-1/4} \int_{\tilde{\sigma}}^\sigma (a'(\sigma))^{-1/4} b'(\sigma) d\sigma$$

where $\tilde{\sigma} = \text{const}$ is the initial value of stress.

3.7.4 Waves in Rods in the Presence of External Friction

Equations of (3.7.9) type also appear in problems of wave propagation in rods in the presence of external friction. Let F be the friction force which acts on the lateral surface of the rod and is counted at a unit of length. Then the equation of motion of the rod will take the form

$$\frac{\partial \sigma}{\partial x} + F = \varrho \frac{\partial^2 u}{\partial t^2}. \tag{3.7.15}$$

There are two possible situations in which this equation reduces to (3.7.9).

a) Linear Elastic Rod, Nonlinear Friction Force. Suppose that the material of the rod obeys the Hooke's law, i.e.,

$$\sigma = E\varepsilon$$

($\varepsilon = \partial u/\partial x$) and the friction force has the form

$$F = -\gamma f\left(\frac{\partial u}{\partial t}\right), \quad 0 < \gamma \ll 1.$$

Then (3.7.15) will evidently take the form

$$E \frac{\partial^2 u}{\partial x^2} - \gamma f\left(\frac{\partial u}{\partial t}\right) = \varrho \frac{\partial^2 u}{\partial t^2}.$$

By differentiating this equation with respect to t and setting $v = \partial u/\partial t$, we arrive at the equation for velocity which belongs to the (3.7.9) type:

$$\frac{\partial^2 v}{\partial t^2} + \frac{\gamma}{\varrho} \frac{\partial f(v)}{\partial t} - \frac{E}{\varrho} \frac{\partial^2 v}{\partial x^2} = 0. \tag{3.7.16}$$

b) Nonlinear Elastic Rod, Linear Friction Force. Let the nonlinear consitutive equation

$$\varepsilon = a(\sigma)$$

hold for the material of the rod, and let the friction force F depend on the velocity of the elements of the rod in a linear way:

$$F = -k\gamma \frac{\partial u}{\partial t}; \quad k > 0, \quad 0 < \gamma \ll 1.$$

Inserting the above expression for F into (3.7.15), we obtain

$$\frac{\partial \sigma}{\partial x} - k\gamma \frac{\partial u}{\partial t} = \varrho \frac{\partial^2 u}{\partial t^2}.$$

Differentiating this equation with respect to x yields

$$\frac{\partial^2 \sigma}{\partial x^2} - k\gamma \frac{\partial \varepsilon}{\partial t} = \varrho \frac{\partial^2 \varepsilon}{\partial t^2}$$

whence, in accordance with the constitutive equation, we finally have

$$\frac{\partial^2 a(\sigma)}{\partial t^2} + \frac{k\gamma}{\varrho}\frac{\partial a(\sigma)}{\partial t} - \frac{1}{\varrho}\frac{\partial^2 \sigma}{\partial x^2} = 0 . \qquad (3.7.17)$$

Thus, we again have arrived at an equation of the (3.7.9) type.

Remark. An interesting theory describing wave propagation in elastic rods in case of external dry friction (where the friction force equals $-k$ sign $\partial u/\partial t$, $k > 0$) is given in [3.10,11]. For some further results concerning hereditary case see [3.12].

3.8 Nonstationary Waves for an Exponential Memory Function

3.8.1 Formulation of the Problem

Let us return to the constitutive equation (3.7.1). Assume now that the hereditary kernel $\Phi(t)$, introduced in (3.7.5), has the form

$$\Phi(t) = k e^{-kt}, \quad k > 0 .$$

Then from Theorem 3.7.2 it follows that stress waves travelling to the right can be described (in the domain of smoothness) by the equation

$$\sqrt{\varrho a'(\sigma)}\,\frac{\partial \sigma}{\partial t} + \frac{\partial \sigma}{\partial x} - k \int_{-\infty}^{t} e^{-k(t-\tau)} \frac{\partial \sigma(\tau, x)}{\partial x}\, d\tau = 0 . \qquad (3.8.1)$$

We shall consider the usual problem

$$\begin{aligned} \sigma &= 0 \quad \text{for} \quad x > 0, \ t \leq 0, \\ \sigma(t, 0) &= \sigma_0(t) \end{aligned} \qquad (3.8.2)$$

where $\sigma_0(t) = 0$ for $t \leq 0$ and for $t \geq T > 0$. The function $\sigma_0(t)$ is assumed to be smooth and distinct from zero on the interval $(0, T)$.

Below we shall demonstrate the way one can obtain the exact solution of (3.8.1,2) in the domain of smoothness.

3.8.2 Derivation of Single–Wave Differential Equation

First of all, since the kernel of the convolution operator in (3.8.1) is exponential, we can use a standard method to eliminate the integral summand in (3.8.1) and reduce (3.8.1) to a second order differential equation.

In fact, differentiating (3.8.1) with respect to t, we have

$$\frac{\partial}{\partial t}\left(\sqrt{\varrho a'(\sigma)}\,\frac{\partial \sigma}{\partial t}\right) + \frac{\partial}{\partial t}\frac{\partial \sigma}{\partial x} - k\frac{\partial \sigma}{\partial x}$$
$$+ k^2 \int_{-\infty}^{t} e^{-k(t-\tau)}\frac{\partial \sigma(\tau, x)}{\partial x}\,d\tau = 0. \qquad (3.8.3)$$

Let us now multiply (3.8.1) by k and add the result to (3.8.3) (obviously, the integral summands as well as the terms containing $\partial \sigma/\partial x$ will vanish as a result of this operation):

$$\frac{\partial}{\partial t}\left(\sqrt{\varrho a'(\sigma)}\,\frac{\partial \sigma}{\partial t}\right) + \frac{\partial}{\partial t}\frac{\partial \sigma}{\partial x} + k\sqrt{\varrho a'(\sigma)}\,\frac{\partial \sigma}{\partial t} = 0.$$

The previous equation evidently can be rewritten in the form

$$\frac{\partial}{\partial t}\left\{\sqrt{\varrho a'(\sigma)}\,\frac{\partial \sigma}{\partial t} + \frac{\partial \sigma}{\partial x} + k\int_0^{\sigma}\sqrt{\varrho a'(\sigma)}\,d\sigma\right\} = 0 \qquad (3.8.4)$$

whence it follows that

$$\sqrt{\varrho a'(\sigma)}\,\frac{\partial \sigma}{\partial t} + \frac{\partial \sigma}{\partial x} + k\int_0^{\sigma}\sqrt{\varrho a'(\sigma)}\,d\sigma = f(x).$$

But since $\sigma = 0$ before the arrival of the wave, then

$$f(x) \equiv 0.$$

Thus, we have arrived at an exact first order differential equation

$$\frac{\partial}{\partial t}\int_0^{\sigma}\sqrt{\varrho a'(\sigma)}\,d\sigma + \frac{\partial \sigma}{\partial x} + k\int_0^{\sigma}\sqrt{\varrho a'(\sigma)}\,d\sigma = 0. \qquad (3.8.5)$$

3.8.3 The Analytic Solution in a Smoothness Domain

It is clear that one can solve (3.8.5) directly by the method of characteristics. However, we can somewhat simplify our formulas by introducing the following change of the unknown function:

$$w = \int_0^{\sigma}\sqrt{\varrho a'(\sigma)}\,d\sigma. \qquad (3.8.6)$$

The inverse relation will be written as

$$\sigma = \varphi(w). \qquad (3.8.7)$$

3.8 Nonstationary Waves for an Exponential Memory Function

Then (3.8.5) will obviously take the form

$$\frac{\partial w}{\partial t} + \varphi'(w)\frac{\partial w}{\partial x} + kw = 0 . \tag{3.8.8}$$

The boundary condition for this equation follows from (3.8.2,6)

$$w(t,0) = \int_0^{\sigma_0(t)} \sqrt{\varrho a'(\sigma)}\, d\sigma \equiv w_0(t) \tag{3.8.9}$$

(it is clear that $w_0(t) = 0$ for $t \leq 0$ and for $t \geq T > 0$ and that $w_0(t) \neq 0$ for $0 < t < T$).

One can see from (3.8.8) that this time it is more convenient to write the equations of characteristics in the following form:

$$\frac{dx}{dt} = \varphi'(w) , \tag{3.8.10}$$

$$\frac{dw}{dt} = -kw . \tag{3.8.11}$$

Integrating (3.8.11) and taking into account (3.8.9), we obtain

$$w = e^{-k(t-\tau)} w_0(\tau) \tag{3.8.12}$$

(τ denotes the value of t for $x = 0$). Inserting now (3.8.12) into (3.8.10), we finally have

$$x = \int_\tau^t \varphi'\left(e^{-k(y-\tau)} w_0(\tau)\right) dy \tag{3.8.13}$$

where we have taken into consideration that $x = 0$ for $t = \tau$.

Thus, (3.8.12,13) solves (3.8.8,9) in the domain of smoothness. Transition to the solution of the original problem (3.8.1,2) is obviously done by (3.8.7).

3.8.4 Wave Breaking

To determine the wave breaking coordinates, we have to find the envelope of the family of characteristics (3.8.13) which depend on the parameter τ. Since $w_0(\tau) = 0$ for $\tau \leq 0$ and $\tau \geq T$, it is clear from (3.8.13) that the characteristics corresponding to the mentioned values of τ are parallel straight lines. Therefore, in seeking the envelope of the family (3.8.13), we can restrict ourselves to the values $0 < \tau < T$.

Let us introduce a change of the variable of integration in (3.8.13) by setting

$$z = e^{-k(y-\tau)} w_0(\tau), \quad 0 < \tau < T . \tag{3.8.14}$$

Hence,

$$dz = -k e^{-k(y-\tau)} w_0(\tau) dy ,$$

i.e.,
$$dy = -\frac{1}{k}\frac{dz}{z}.$$

Furthermore, for $y = \tau$ it is obvious that
$$z = w_0(\tau)$$
and for $y = t$ it is also obvious that
$$z = e^{-k(t-\tau)}w_0(\tau).$$

Thus, the equation of the family of characteristics (3.8.13) (for $0 < \tau < T$) can be rewritten as
$$x = \frac{1}{k}\int_{\exp[-k(t-\tau)]w_0(\tau)}^{w_0(\tau)} \frac{\varphi'(z)}{z}\,dz. \tag{3.8.15}$$

Now, to find the envelope of the family of characteristics under consideration, let us differentiate (3.8.15) with respect to the parameter τ.
We have
$$0 = \frac{1}{k}w_0'(\tau)\frac{\varphi'(w_0(\tau))}{w_0(\tau)} - \frac{1}{k}e^{-kt}\left(e^{k\tau}w_0(\tau)\right)'$$
$$\times \frac{\varphi'\left(e^{-kt}e^{k\tau}w_0(\tau)\right)}{e^{-kt}e^{k\tau}w_0(\tau)}, \quad 0 < \tau < T. \tag{3.8.16}$$

The equations (3.8.15,16) are the ones that determine (in a parametric form) the envelope of the family of characteritics considered.

As we know, the wave breaking time in our problem is the minimal positive value of t–coordinate of the points of the envelope. To determine the minimal value only one equation, (3.8.16), must be used. Let us rewrite (3.8.16) in a more convenient form:
$$w_0'(\tau)\varphi'(w_0(\tau)) = \left(kw_0(\tau) + w_0'(\tau)\right)\varphi'\left(e^{-kt}e^{k\tau}w_0(\tau)\right)$$
whence
$$t = -\frac{1}{k}\ln\left\{\frac{e^{-k\tau}}{w_0(\tau)}(\varphi')^{-1}\left(\frac{w_0'(\tau)\varphi'(w_0(\tau))}{kw_0(\tau) + w_0'(\tau)}\right)\right\}. \tag{3.8.17}$$

The minimal value $t = t_0$ which can be found from (3.8.17) and which corresponds to some $\tau = \tau^*$ will be the desired wave breaking time. The appropriate distance $x = x_0$ can be calculated by (3.8.15), where one must replace t by t_0 and τ by its value τ^*. The absence of positive values of t among those given by (3.8.17) implies that wave breaking does not occur at all in the problem under consideration.

Problem. What changes will the above calculation undergo for the function $\sigma_0(t)$ with an isolated zero within $0 < t < T$?

3.8.5 Case of Small Amplitudes. Asymptotic Analysis of the Shock–Wave

In the case of small amplitudes we shall obtain explicit asymptotic formulas for the wave breaking coordinates in the problem considered. In addition, it is possible to give a qualitative description of the stress wave after formation of the shock-wave front.

Therefore, suppose for simplicity

$$\int_0^\sigma \sqrt{\varrho a'(\sigma)}\, d\sigma = A\sigma + B\sigma^2\,; \quad A > 0\,. \tag{3.8.18}$$

Then (3.8.5) [which follows, as we know, from (3.8.1)] will take the form

$$A\frac{\partial \sigma}{\partial t} + B\frac{\partial \sigma^2}{\partial t} + \frac{\partial \sigma}{\partial x} + kA\sigma + kB\sigma^2 = 0\,.$$

Neglecting here the last summand, we finally arrive at

$$A\frac{\partial \sigma}{\partial t} + B\frac{\partial \sigma^2}{\partial t} + \frac{\partial \sigma}{\partial x} + C\sigma = 0 \tag{3.8.19}$$

where $C \equiv kA$. It is clear that $C > 0$ since k and A are positive.

Now, in (3.8.19) we shall introduce a change of the unknown function

$$\sigma = \frac{f}{2B}\,e^{-Cx} \tag{3.8.20}$$

and a change of the independent variables

$$x_1 = \frac{1 - e^{-Cx}}{C}\,, \quad t_1 = t - Ax\,. \tag{3.8.21}$$

Then (3.8.19) will take the form

$$f\frac{\partial f}{\partial t_1} + \frac{\partial f}{\partial x_1} = 0\,. \tag{3.8.22}$$

Furthermore, it is obvious that, under the mapping (3.8.21), the semi-plane $x \geq 0$ will go to the band

$$0 \leq x_1 < \frac{1}{C}\,; \tag{3.8.23}$$

it is also clear that t_1–axis will occupy the place of t–axis, the positive direction being preserved.

Note now that since we have extended the boundary function $\sigma_0(t)$ by zero to the semi-axis $t < 0$, the first of the conditions of (3.8.2) proves to be superfluous. The second of the conditions of (3.8.2) takes the following form in new coordinates:

$$f(t_1, 0) = 2B\sigma_0(t_1)\,. \tag{3.8.24}$$

Thus, we have arrived at (3.8.22,24) whose solution was studied in detail in Sect. 1.2.

We want to emphasize the fact that we have to study this solution only in the band (3.8.23).

In Sect. 3.8.4 we have already derived formulas for the wave breaking coordinates. However, under the approach employing the smallness of wave amplitude these formulas can be considerably simplified. Indeed, by virtue of the results of Sect. 1.2.4 the x_1–coordinate of the wave breaking equals

$$-\frac{1}{2B\sigma_0'(\tau^*)}$$

where τ^* is the point at which the quantity $-2B\sigma_0'(\tau)$ achieves its maximum. This maximum will obviously be positive under our assumptions concerning the function σ_0.

Hence [see (3.8.21)] the shock appears for

$$x = x_0 = -\frac{1}{C}\ln\left(1 + \frac{C}{2B\sigma_0'(\tau^*)}\right). \tag{3.8.25}$$

Using the equations of characteristics, one can easily find the moment t_0 of the wave breaking. Furthermore, it is evident that the condition

$$1 + \frac{C}{2B\sigma_0'(\tau^*)} > 0 \tag{3.8.26}$$

must be satisfied; otherwise wave breaking will be impossible for real x. Thus, for small amplitudes the appearance of wave breaking depends only on the maximal steepness of the boundary profile and does not depend on other features of the profile.

Now assuming that (3.8.26) is satisfied, let us turn to the study of the emerging shock-wave. For simplicity we shall suppose that the boundary function σ_0 has the unique extremum. Thus, we consider the overtaking of "a single hump".

It is easy to check that in the new variables f, t_1, x_1 the condition on the shock (3.7.7) asymptotically assumes the form

$$V = \frac{[f]}{[f^2/2]} \tag{3.8.27}$$

where V is the shock-wave front velocity in coordinates t_1, x_1. Therefore, because of the results of Sect. 1.2, in the problem under consideration the shock-wave front can easily be constructed by the method of equal areas. Let

$$t_1 = S(x_1) \tag{3.8.28}$$

be the equation of the shock wave front in coordinates t_1, x_1. As we know from Sect. 1.2, the function $S(x_1)$ is defined and continuous everywhere to the right of the point at which wave breaking occurs, i.e., for

$$x_1 > -\frac{1}{2B\sigma_0'(\tau^*)}. \qquad (3.8.29)$$

[We are, however, interested in the values of $S(x_1)$ only in the interval (3.8.23).] This simple observation allows us to give a qualitative description of the shock-wave front in the original coordinates. In fact, inserting t_1, x_1, given by (3.8.21), into (3.8.28) we have

$$t - Ax = S\left(\frac{1 - e^{-Cx}}{C}\right), \quad x > x_0$$

whence, for x large enough,

$$t \sim Ax + S\left(\frac{1}{C}\right). \qquad (3.8.30)$$

As to the stress shock on the front, its value will decrease exponentially (as $x \to -\infty$) by virtue of

$$[\sigma] = \frac{[f]}{2B} e^{-Cx} \qquad (3.8.31)$$

which follows from (3.8.20). In fact, for all finite x satisfying the inequality (3.8.29) the shock $[f]$ varies continuously and is distinct from zero (Sect. 1.2.5). In particular, for $x_1 \to 1/C$ the value of $[f]$ tends to some nonzero constant. But $x_1 \to 1/C$ corresponds to $x \to \infty$. This proves that

$$\frac{[f]}{2B} \to \text{const} \neq 0 \quad \text{as} \quad x \to \infty$$

whence it follows that the shock $[\sigma]$ is an exponentially decreasing function of x.

Thus, in the case of small amplitudes and exponential kernels of a special form we have succeeded in describing qualitatively the behaviour of a shock-wave front which emerges at the moment of breaking of a solitary wave. However, in accordance with the previous results (Sect. 3.4) shock-wave fronts may appear practically for any reasonable hereditary kernels. Carrying out an adequate analysis of the emerging shock-wave fronts is an interesting open problem.

3.9 Reflection of a Wave from the Boundary Between Linear Elastic and Nonlinear Hereditary Media

3.9.1 Formulation of the Boundary Value Problem

Theorem 3.7.2 concerning the exact factorization of a nonlinear wave equation with memory enables us to reduce the following problem to an integral equation.

Let us consider a compound rod located on the semi-axis $x \geq 0$. Suppose the part $0 \leq x < l$ of the rod under consideration is linear elastic with density

ϱ_1 = const and Young's modulus E_1 = const. Thus, stress in this part of the rod satisfies the wave equation

$$\frac{\partial^2 \sigma}{\partial t^2} - \frac{E_1}{\varrho_1} \frac{\partial^2 \sigma}{\partial x^2} = 0, \quad 0 < x < l. \tag{3.9.1}$$

Now suppose that the part $x > l$ of the rod under consideration has density ϱ_2 = const and assume that the corresponding constitutive equation has the form (3.7.1). Therefore, waves travelling to the right are described by (3.7.6):

$$\sqrt{\varrho_2 a'(\sigma)} \frac{\partial \sigma}{\partial t} + (1 - \Phi^*) \frac{\partial \sigma}{\partial x} = 0, \quad x > l. \tag{3.9.2}$$

It is evident that on the boundary $x = l$ the following two conditions must hold:

$$\sigma^- = \sigma^+ \tag{3.9.3}$$

(the equality of stresses) and

$$\frac{1}{\varrho_1} \left(\frac{\partial \sigma}{\partial x} \right)^- = \frac{1}{\varrho_2} \left(\frac{\partial \sigma}{\partial x} \right)^+ \tag{3.9.4}$$

(the equality of accelerations of material elements). The indices "-" and "+" denote here the limiting values of the quantities under consideration at the approach to the boundary from the left and right, respectively.

Suppose, finally, that

$$\begin{aligned} \sigma = 0 \quad &\text{for} \quad x > 0, \ t \leq 0, \\ \sigma(t, 0) &= \sigma_0(t) \end{aligned} \tag{3.9.5}$$

where $\sigma_0(t) = 0$ for $t < 0$.

Our aim is to find the stress wave that appears in the linear elastic part of the rod as a result of repeated reflections of the incident wave from the boundaries $x = 0$ and $x = l$.

The analysis given below will be limited by the time interval during which no shock-wave is formed in the right-hand part of the rod. Clearly, (3.9.2) will be valid during this time interval.

3.9.2 Reduction of the Problem to an Integro-Functional Equation

First of all, from (3.9.1,5) it easily follows that for $0 \leq x < l$ the solution of the problem under consideration can be represented as

$$\sigma = \sigma_0 \left(t - \frac{x}{c_1} \right) - \sigma_1 \left(t - \frac{x}{c_1} \right) + \sigma_1 \left(t + \frac{x}{c_1} \right), \quad c_1 = \sqrt{\frac{E_1}{\varrho_1}} \tag{3.9.6}$$

3.9 Reflection of a Wave

where σ_1 is a function to be determined. Since the reflected wave first appears for $x = l$ at the moment $t = l/c_1$, the function σ_1 must evidently satisfy the condition

$$\sigma_1(z) = 0 \quad \text{for} \quad z < 2l/c_1. \tag{3.9.7}$$

Now (3.9.3) can be rewritten in the form

$$\sigma_0\left(t - \frac{l}{c_1}\right) - \sigma_1\left(t - \frac{l}{c_1}\right) + \sigma_1\left(t + \frac{l}{c_1}\right) = \sigma^+. \tag{3.9.8}$$

Furthermore, from (3.9.2) we have

$$\left(\frac{\partial \sigma}{\partial x}\right)^+ = -\frac{1}{1 - \Phi^*}\left(\sqrt{\varrho_2 a'(\sigma^+)}\left(\frac{\partial \sigma}{\partial t}\right)^+\right)$$

$$\equiv -\frac{1}{1 - \Phi^*}\left(\sqrt{\varrho_2 a'(\sigma^+)}\,\frac{d\sigma^+}{dt}\right).$$

From the last relation and (3.9.6) it evidently follows that (3.9.4) can be rewritten in the form

$$\frac{1}{\varrho_1 c_1}\left\{-\sigma_0'\left(t - \frac{l}{c_1}\right) + \sigma_1'\left(t - \frac{l}{c_1}\right) + \sigma_1'\left(t + \frac{l}{c_1}\right)\right\}$$

$$= -\frac{1}{1 - \Phi^*}\left(\sqrt{\frac{a'(\sigma^+)}{\varrho_2}}\,\frac{d\sigma^+}{dt}\right). \tag{3.9.9}$$

Equations (3.9.8,9) present a system with respect to the unknown functions σ_1 and σ^+, which we must solve.

Integrating (3.9.9) over t and taking into account the commutativity of integration and convolution, we easily obtain

$$\frac{1}{\varrho_1 c_1}\left\{-\sigma_0\left(t - \frac{l}{c_1}\right) + \sigma_1\left(t - \frac{l}{c_1}\right) + \sigma_1\left(t + \frac{l}{c_1}\right)\right\}$$

$$= -\frac{1}{\sqrt{\varrho_2}(1 - \Phi^*)}\int_0^{\sigma^+} \sqrt{a'(\sigma)}\,d\sigma + \text{const}.$$

After taking an arbitrary t from the interval $(0, l/c_1)$, we obviously have const $= 0$ in the previous equality, which therefore yields

$$(1 - \Phi^*)\left\{\sigma_0\left(t - \frac{l}{c_1}\right) - \sigma_1\left(t - \frac{l}{c_1}\right) - \sigma_1\left(t + \frac{l}{c_1}\right)\right\} = F(\sigma^+) \tag{3.9.10}$$

where

$$F(\sigma^+) \equiv \sqrt{\frac{E_1 \varrho_1}{\varrho_2}}\int_0^{\sigma^+} \sqrt{a'(\sigma)}\,d\sigma.$$

Now, inserting the expression for σ^+ from (3.9.8) into (3.9.10), we obtain the following nonlinear integro-functional equation:

$$(1 - \Phi^*)\left\{\sigma_0\left(t - \frac{l}{c_1}\right) - \sigma_1\left(t - \frac{l}{c_1}\right) - \sigma_1\left(t + \frac{l}{c_1}\right)\right\}$$
$$= F\left(\sigma_0\left(t - \frac{l}{c_1}\right) - \sigma_1\left(t - \frac{l}{c_1}\right) + \sigma_1\left(t + \frac{l}{c_1}\right)\right). \qquad (3.9.11)$$

Here the unknown function σ_1 enters with shifted arguments.

3.9.3 Solution of the Integro-Functional Equation

On the face of it, (3.9.11) seems to be rather complicated, but it can be easily solved by means of a standard technique.

It is clear that by virtue of (3.9.7) it is reasonable to consider (3.9.11) only for $t \geq l/c_1$. At first, suppose that $l/c_1 \leq t < 3l/c_1$. By virtue of (3.9.7) the function $\sigma_1(t - l/c_1)$ becomes identically equal to zero in the indicated half-interval; thus (3.9.11) can be rewritten as

$$(1 - \Phi^*)\left\{\sigma_0\left(t - \frac{l}{c_1}\right) - \sigma_1\left(t + \frac{l}{c_1}\right)\right\}$$
$$= F\left(\sigma_0\left(t - \frac{l}{c_1}\right) + \sigma_1\left(t + \frac{l}{c_1}\right)\right); \quad \frac{l}{c_1} \leq t < \frac{3l}{c_1}.$$

This equation can easily be solved by the method of successive approximations. However, by this token the function $\sigma_1(t - l/c_1)$ will be determined for $3l/c_1 \leq t < 5l/c_1$. Hence, for $3l/c_1 \leq t < 5l/c_1$ the method of successive approximations can be applied anew to solve (3.9.11) with respect to $\sigma_1(t + l/c_1)$ by using the obtained value of $\sigma_1(t - l/c_1)$. Proceeding in this manner, we obtain the desired solution of (3.9.11) for all t.

For details that pertain to the case of small quadratic nonlinearity, see [3.21].

In conclusion we would like to emphasize that the key point in the reduction to the integro-functional equation (3.9.11) was the expression on the boundary of the spatial derivative $(\partial\sigma/\partial x)^+$ by $(\partial\sigma/\partial t)^+ \equiv d\sigma^+/dt$. For this the fact that the motion in the right-hand part of the rod can be exactly described by a single-wave equation for the chosen model of the medium was of essential importance.

The obvious consequence of our results is that the problem of wave propagation in one part of the rod becomes independent of that in the other. Finally, it is clear that if in (3.9.2) $\Phi^* = 0$, then the stress wave in the right-hand part of the rod can analytically be calculated by the method of characteristics.

Problem. Extend the results of Sect. 3.9 to the case where the left-hand part of the rod is linear hereditary elastic.

3.10 The Exactly Factorizable Linear Wave Equation with Memory and a Variable Coefficient

3.10.1 Factorization Theorem

Here we adduce a rather natural generalization of the results of Sect. 2.2.2. Assume that the density of the rod changes in accordance with the law

$$\varrho = (C_1 x + C_2)^{-4/3}$$

(we suppose $C_1 x + C_2 > 0$ in the considered interval of change of x) and stress is related to deformation by the linear hereditary constitutive equation

$$(1 - R^*)\varepsilon = A\sigma; \quad A = \text{const} > 0.$$

Then, as one can easily check, the corresponding wave equation for stress will take the form

$$A \frac{\partial^2 \sigma}{\partial t^2} - (1 - R^*) \frac{\partial}{\partial x}(C_1 x + C_2)^{4/3} \frac{\partial \sigma}{\partial x} = 0. \qquad (3.10.1)$$

Note that the small parameter γ is not presented in this equation.

As we know, the operator R^* is commutative with differentiation and multiplication by a function depending on x. Therefore, taking into account the analogy with the decomposition (2.2.7), one can easily obtain the following.

Theorem 3.10.1. The equation (3.10.1) can be exactly factorized in the following way:

$$\left\{\sqrt{A}\frac{\partial}{\partial t} \mp \sqrt{1 - R^*}\left(\frac{\partial}{\partial x}(C_1 x + C_2)^{2/3} - \frac{C_1}{3}(C_1 x + C_2)^{-1/3}\right)\right\}$$
$$\times \left\{\sqrt{A}\frac{\partial}{\partial t} \pm \sqrt{1 - R^*}\left((C_1 x + C_2)^{2/3}\frac{\partial}{\partial x}\right.\right.$$
$$\left.\left. + \frac{C_1}{3}(C_1 x + C_2)^{-1/3}\right)\right\}\sigma = 0. \qquad (3.10.2)$$

The validity of the decomposition (3.10.2) of (3.10.1) can be demonstrated by direct multiplication.

Remark. A special case of (3.10.2), the factorization under the condition of constant density, was obtained earlier in [3.22].

3.10.2 Solution of the Boundary Value Problem

From (3.10.2) it evidently follows that the wave travelling to the right can be exactly described by the equation

$$\sqrt{A}\frac{\partial \sigma}{\partial t} + \sqrt{1-R^*}\left((C_1 x + C_2)^{2/3}\frac{\partial \sigma}{\partial x} + \frac{C_1}{3}(C_1 x + C_2)^{-1/3}\sigma\right) = 0 \,.\quad(3.10.3)$$

Furthermore, let

$$\sigma = 0 \quad \text{for} \quad x > 0, \quad t \le 0,$$
$$\sigma(t,0) = \sigma_0(t) \qquad\qquad\qquad\qquad\qquad (3.10.4)$$

where $\sigma_0(t) = 0$ for $t < 0$.

By applying the Laplace transform $L_{t\to p}$, one easily obtains the solution of (3.10.3,4) in the form

$$\sigma(t,x) = \left(\frac{C_2}{C_1 x + C_2}\right)^{1/3} L^{-1}_{p\to t}\bar{\sigma}_0(p)$$

$$\times \exp\left\{-\frac{p}{\sqrt{1-\bar{R}(p)}}\frac{3\sqrt{A}\left((C_1 x + C_2)^{1/3} - C_2^{1/3}\right)}{C_1}\right\} \quad (3.10.5)$$

where

$$\bar{\sigma}_0(p) = L_{t\to p}\sigma_0(t)\,, \quad \bar{R}(p) = L_{t\to p}R(t)\,.$$

Here we shall not carry out the asymptotic analysis of the solution obtained, because in [3.3] such an analysis (using Tauberian methods) for expressions similar to (3.10.5) was given.

References

Chapter 1

1.1 B.L. Rozhdestwensky, N.N. Yanenko: *Systems of Quasilinear Equations* (Nauka, Moscow 1968) [in Russian]
1.2 G.B. Whitham: *Linear and Nonlinear Waves* (Wiley, New York 1974)
1.3 R. Courant, K.O. Friedrichs: *Supersonic Flow and Shock Waves* (Interscience, New York 1948)
1.4 O.A. Oleinik: Usp. Mat. Nauk **3**, 3 (1957)
1.5 P.D. Lax: Comm. Pure Appl. Math. **10**, 537 (1957)
1.6 D.R. Bland: *Nonlinear Dynamic Elasticity* (Waltham, Massachusetts 1969)
1.7 G.I. Barenblatt: Prikl. Mat. Mekh. **17**, 455 (1953)
1.8 M.A. Grinfeld: Prikl. Mat. Mekh. **42**, 883 (1978)
1.9 J. Lighthill: *Waves in Fluids* (Cambridge University Press, Cambridge 1978)
1.10 A.A. Lokshin: Izv. Akad. Nauk SSSR MTT **6**, 104 (1985)
1.11 S. Earnshaw: Phil. Trans. Roy. Soc. London **150**, 133 (1860)
1.12 A.A. Lokshin: Izv. Akad. Nauk SSSR MTT **2**, 134 (1986)
1.13 O.A. Oleinik: Usp. Mat. Nauk **6**, 169 (1957)

Chapter 2

2.1 A.A. Lokshin, E.A. Sagomonyan: Izv. Akad. Nauk SSSR MTT **1**, 95 (1987)
2.2 V.E. Raspopov, V.P. Shapeev: Chisl. Met. Mech. Splosh. Sredy **1**, 2, 76 (1970)
2.3 A.F. Sidorov, V.P. Shapeev, N.N. Yanenko: *The Method of Differential Dependences* (Nauka, Novosibirsk 1984) [in Russian]
2.4 B.L. Rozhdestwensky, N.N. Yanenko: *Systems of Quasi-Linear Equations* (Nauka, Moscow 1968) [in Russian]

Chapter 3

3.1 J.N. Rabotnov: *Elements of Hereditary Mechanics of Solids* (Nauka, Moscow 1977) [in Russian]
3.2 A.J. Sagomonyan: *Stress Waves in Continuous Media* (Moscow University Press, Moscow 1985) [in Russian]
3.3 A.A. Lokshin, J.V. Suvorova: *Mathematical Theory of Wave Propagation in Media with Memory* (Moscow University Press, Moscow 1982) [in Russian]
3.4 E.I. Shemyakin: Dokl. Akad. Nauk **104**, 193 (1955)
3.5 M.J. Kelbert, I.A. Chaban: Izv. Akad. Nauk SSSR MZhG **4**, 164 (1986)
3.6 V.E. Nakoryakov, B.G. Pokusaev, I.R. Shreiber: *Wave Propagation Processes in Gas- and Vapour-Liquid Media* (Institute of Thermophysics, Novosibirsk 1983) [in Russian]
3.7 A.A. Lokshin: Izv. Akad. Nauk SSSR MTT **6**, 104 (1985)
3.8 O.V. Rudenko, S.I. Soluyan: *Theoretical Basis of Nonlinear Acoustics* (Nauka, Moscow 1975) [in Russian]

3.9 J.K. Engelbrecht, U.K. Nigul: *Nonlinear Deformation Waves* (Nauka, Moscow 1981) [in Russian]
3.10 L.V. Nikitin: Izv. Akad. Nauk SSSR MTT **2**, 166 (1967)
3.11 L.V. Nikitin: Izv. Akad. Nauk SSSR MTT **6**, 137 (1978)
3.12 O.S. Vinogradova, A.A. Lokshin, V.E. Rok: Izv. Akad. Nauk SSSR MTT **1**, 152 (1989)
3.13 A.A. Lokshin: Prikl. Mat. Mekh. **5**, 880 (1987)
3.14 U.K. Nigul: *Nonlinear Acousto-Diagnostics* (Sudostroyenie, Leningrad 1981) [in Russian]
3.15 U.K. Nigul, A.S. Stulov: in *Contributions to Nonlinear Mechanics of Continuous Media*, ed. by N.D. Veksler (Valgus, Tallinn 1985) pp.95-101
3.16 L.D. Landau: Prikl. Mat. Mekh. **9**, 496 (1945)
3.17 G.B. Whitham: *Linear and Nonlinear Waves* (Wiley, New York 1974)
3.18 A.A. Lokshin, E.A. Sagomonyan: Izv. Akad. Nauk SSSR MTT **6**, 116 (1989)
3.19 A.A. Lokshin, M.A. Itskovits: Int. J. Nonlin. Mech. **23**, 125 (1988)
3.20 U.K. Nigul: in *Contributions to Nonlinear Mechanics of Continuous Media*, ed. by N.D. Veksler (Valgus, Tallinn 1985) pp.147-160
3.21 A.A. Lokshin: Izv. Akad. Nauk SSSR MTT **2**, 134 (1986)
3.22 A.A. Lokshin, V.E. Rok: Dokl. Akad. Nauk SSSR **239**, 1305 (1978)

Subject Index

analog of Landau-Whitham formula 93
analytic
   – solution 106
   – – in a smoothness domain 106
application of Euler's method 31
asymptotic
   – analysis of shock-wave 109
   – factorization 33, 37, 42, 52
   – – of nonlinear wave equation 42, 58
   – – – – with memory 58
   – – – – with variable coefficients 42
   – – theorem 101

boundary value problem 9, 20, 46, 98, 111
   – – –, formulation of 37
   – – – in terms of displacement 20, 98
   – – –, solution of 9, 116
   – – –, – single-wave 38

conditions
   – on shock 7
   – on shock for stress wave 61
   – of asymptotic factorization 37, 46
   – on strong shock 4
construction of $g(\sigma)$ 25
continuous stationary-profile waves 63
curve of maximums 49

displacement 20

Earnshaw's theorem 18
   –, generalization of 19
elastic media 111
equal areas 12
   – –, principle of 12
equation
   – for maximal amplitude 48
   – of motion 1
Euler's method 31
exact factorization 15, 39
   – –, linear case 40
   – –, nonlinear case 39

exponential
   – kernel 71, 77, 86
   – memory function 105
external friction 104

factorization
   –, linear 17
   –, nonlinear 17
   – theorem 16, 115
   – –, asymptotic 101
   – –, exact 99
   – – for deformation wave equation 18
   – –, for systems with memory 99
   – –, further 99
formulation of boundary value
   problem 37, 46, 47
friction force, linear 104
function $g(\sigma)$ 25

hereditary media 111
   – elasticity 55
   – –, linear equations of 55
   – –, nonlinear equations of 56
homogeneous
   – integral Volterra equations 63
   – – – –, solutions of 63
   – nonlinear rod 1
hyperbolic equations 7
   – of first order 7
   – – –, nonlinear 7

integral equation 80, 81
   – – for $g(\sigma)$ 24
   – – – generating transformation 24
   – – – stationary-profile wave 80
   – –, estimate of solution 82
   – –, solution of 82
   – –, Volterra 63
   – –, – –, self-coordinated 80
integrals of solution 8
   –, constancy of 8

Subject Index

integro-functional equation 112
   – – –, solution of 114

kernel
   –, exponential 71, 77, 86
   –, oscillatory 72
   –, power 85

Landau-Whitham formula 93
   – – –, analog of 93
linear
   – friction force 104
   – wave equation with variable coefficient 115

maximal amplitude
   – –, equation for 48
   – – for stress wave 48
media
   – with memory 55
   – – –, nonlinear wave in 55
   –, linear elastic 111
   –, nonlinear hereditary 111
memory, linear wave equation with 115
method of characteristics 9

nonlinear
   – elastic rod 104
   – rod 1
   – short waves 33
   – wave equation 15
   – – – with variable coefficient 33
   – – – – constant coefficient 15
   – waves in homogeneous media 1
   – hyperbolic equations of first order 7
nonstationary waves 93

ordinary differential equation 14
   – – – for shock 14
oscillatory kernel 72
   – –, complicated 72
   – –, simplest 72

power kernel 85
pre-stressed medium 75, 87
principle of equal areas 12, 30
propagation of stress wave 50
   – – – in homogeneous elastic rod 50

quadratic nonlinearity 58

reflection of wave 111
   – – from boundary 111

Riemann characteristics 2
   – invariants 2
rod 1
   –, homogeneous elastic 50
   –, – nonlinear 1
   –, nonlinear elastic 104
Rok's method 92

shock 4, 14, 61
   – condition 7, 23
   – for stress wave 61
   – – – –, condition on 61
   – propagating into undisturbed domain 14
   –, strong 4, 7
   –, weak 7
shock-wave 22, 33, 80, 83, 109
   – in a simple system 22, 30
   –, asymptotic analysis 109
short waves
   – – of finite amplitude 33
   – – – – in inhomogeneous media 33
single-wave
   – differential equation 106
   – equation 23, 24, 53, 60, 96
   – – and shock condition 23
   – – for deformation 96
   – – for displacement 97
   – – for homogeneous elastic rod 53
   – solution 38
   – – of boundary value problem 38
solution
   – boundary value problem 9
   – integral equation, estimate of 64
stability condition for strong shock 5
stationary-profile
   – shock-wave 80
   – –, existence of 83
   – wave 63
   – –, existence of 66
   – –, –, general case of 69
   – –, –, special case of 66
   – –, integral equation for 80
stress wave 50
strong shock 4, 7
   – –, conditions on 4
   – –, –, stability 5

undisturbed medium 80
   – –, waves in 80

wave(s)
- breaking 11, 107
- equation 3
- - for stress 16
- - with a variable coefficient 33
- in pre-stressed medium 75, 87
- in rods 104
- - - with external friction 104
- in undisturbed medium 63, 80
- tending to a stationary profile 89

Printing: Weihert-Druck GmbH, Darmstadt
Binding: Theo Gansert Buchbinderei GmbH, Weinheim

D. Park, Williams College, Williamstown, MA

Classical Dynamics and Its Quantum Analogues

2nd enl. and updated ed. 1990. IX, 333 pp. 101 figs.
Hardcover ISBN 3-540-51398-1

The primary purpose of this textbook is to introduce students to the principles of classical dynamics of particles, rigid bodies, and continuous systems while showing their relevance to subjects of contemporary interest. Two of these subjects are quantum mechanics and general relativity. The book shows in many examples the relations between quantum and classical mechanics and uses classical methods to derive most of the observational tests of general relativity. A third area of current interest is in nonlinear systems, and there are discussions of instability and of the geometrical methods used to study chaotic behaviour. In the belief that it is most important at this stage of a student's education to develop clear conceptual understanding, the mathematics is for the most part kept rather simple and traditional. This book devotes some space to important transitions in dynamics: the development of analytical methods in the 18th century and the invention of quantum mechanics.

A. Hasegawa, AT&T Bell Laboratories, Murray Hill, NJ

Optical Solitons in Fibers

2nd enl. ed. 1990. XII, 79 pp. 25 figs. Softcover ISBN 3-540-51747-2

Already after six months high demand made a new edition of this textbook necessary. The most recent developments associated with two topical and very important theoretical and practical subjects are combined: **Solitons** as analytical solutions of nonlinear partial differential equations and as lossless signals in dielectric **fibers**. The practical implications point towards technological advances allowing for an economic and undistorted propagation of signals revolutionizing telecommunications. Starting from an elementary level readily accessible to undergraduates, this pioneer in the field provides a clear and up-to-date exposition of the prominent aspects of the theoretical background and most recent experimental results in this new and rapidly evolving branch of science. This well-written book makes not just easy reading for the researcher but also for the interested physicist, mathematician, and engineer. It is well suited for undergraduate or graduate lecture courses.

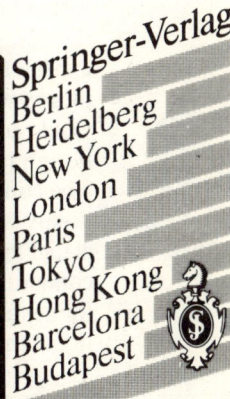

Springer-Verlag
Berlin
Heidelberg
New York
London
Paris
Tokyo
Hong Kong
Barcelona
Budapest

B. N. Zakhariev, Moscow; A. A. Suzko, Minsk

Direct and Inverse Problems

Potentials in Quantum Scattering

1990. XIII, 223 pp. 42 figs. Softcover ISBN 3-540-52484-3

This textbook can almost be viewed as a "how-to" manual for solving quantum inverse problems, that is, for deriving the potential from spectra and/or scattering data. The formal exposition of inverse methods is paralleled by a discussion of the direct problem. In part differential and finite-difference equations are presented side by side. A variety of solution methods is presented. Their common features and (dis)advantages are analyzed. To foster a better understanding, the physical meaning of the mathematical quantities are discussed in detail.

Wave confinement in continuum bound states, resonance and collective tunneling, and the spectral and phase equivalence of various interactions are some of the physical problems covered.

R. M. Dreizler, University of Frankfurt;
E. K. U. Gross, University of Würzburg

Density Functional Theory

An Approach to the Quantum Many-Body Problem

1990. XI, 302 pp. 18 figs. Hardcover ISBN 3-540-51993-9

Density Functional Theory is a rapidly developing branch of many-particle physics that has found applications in atomic, molecular, solid state and nuclear physics. This book describes the conceptual framework of density functional theory and discusses in detail the derivation of explicit functionals from first principles as well as their application to Coulomb systems. Both non-relativistic and relativistic systems are treated. The connection of density functional theory with other many-body methods is highlighted. The presentation is self-contained; the book is thus well suited for a graduate course on density functional theory.

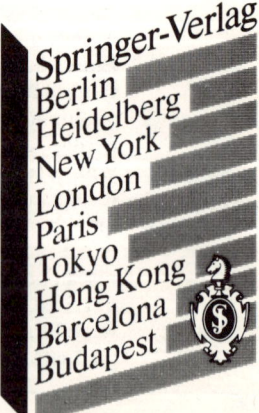

Springer-Verlag
Berlin
Heidelberg
New York
London
Paris
Tokyo
Hong Kong
Barcelona
Budapest

Research Reports in Physics

The categories of camera-ready manuscripts (e.g., written in T_EX; preferably both hard and soft copy) considered for publication in the **Research Reports** include:

1. Reports of meetings of particular interest that are devoted to a single topic (provided that the camera-ready manuscript is received within four weeks of the meeting's close!).
2. Preliminary drafts of original papers and monographs.
3. Seminar notes on topics of current interest.
4. Reviews of new fields.

Should a manuscript appear better suited to another series, consent will be sought from the author for its transfer to the other series.

Research Reports in Physics are divided into numerous subseries, e.g., nonlinear dynamics or nuclear and particle physics. Besides covering material of general interest, the series provides the possibility for topics that are too specialized or controversial to be published within the traditional avenues. The small print runs make a consistent price structure impossible and will sometimes have to presuppose a financial contribution from the author (or a sponsor). In particular, in the case of proceedings the organizers are expected to place a bulk order and/or provide some funding.

Within **Research Reports** the timeliness of a manuscript is more important than its form, which may be unfinished or tentative. Thus in some instances, proofs may be merely outlined and results presented that will be published in full elsewhere later. Since the manuscripts are directly reproduced, the responsibility for form and content is mainly the author's.

Springer-Verlag
Berlin
Heidelberg
New York
London
Paris
Tokyo
Hong Kong
Barcelona
Budapest

☐ Heidelberger Platz 3, W-1000 Berlin 33, F. R. Germany ☐ 175 Fifth Ave., New York, NY 10010, USA ☐ 8 Alexandra Rd., London SW19 7JZ, England ☐ 26, rue des Carmes, F-75005 Paris, France ☐ 37-3, Hongo 3-chome, Bunkyo-ku, Tokyo 113, Japan ☐ Room 701, Mirror Tower, 61 Mody Road, Tslmshatsul, Kowloon, Hong Kong ☐ Avinguda Diagonal, 468-4° C, E-08006 Barcelona, Spain

Research Reports in Physics

Manuscripts should be no less than 100 and no more than 400 pages in length. They are reproduced by a photographic process and must therefore be typed with extreme care. Corrections to the typescript should be made by pasting in the new text or painting out errors with white correction fluid. The typescript is reduced slightly in size during reproduction; the text on every page has to be kept within a frame of 16 × 25.4 cm (6 5/16 × 10 inches). On request, the publisher will supply special stationery with the typing area outlined.

Editors or authors (of complete volumes) receive 5 complimentary copies and are free to use parts of the material in later publications.

All manuscripts, including proceedings, must contain a subject index. In the case of multi-author books and proceedings an index of contributors is also required. Proceedings should also contain a list of participants, with complete addresses.

Our leaflet, *Instructions for the Preparation of Camera-Ready Manuscripts*, and further details are available on request.

Manuscripts (in English) or inquiries should be directed to

Dr. Ernst F. Hefter
Physics Editorial 4
Springer-Verlag, Tiergartenstrasse 17
W-6900 Heidelberg, Fed. Rep. of Germany
(Tel. [0]6221-487495;
Telex 461723;
Telefax [0]6221-413982)

☐ Heidelberger Platz 3, W-1000 Berlin 33, F. R. Germany ☐ 175 Fifth Ave., New York, NY 10010, USA ☐ 8 Alexandra Rd., London SW19 7JZ, England ☐ 26, rue des Carmes, F-75005 Paris, France ☐ 37-3, Hongo 3-chome, Bunkyo-ku, Tokyo 113, Japan ☐ Room 701, Mirror Tower, 61 Mody Road, Tsimshatsui, Kowloon, Hong Kong ☐ Avinguda Diagonal, 468-4" C, E-08006 Barcelona, Spain